HOW TO B [illegible] POWER TUNE

SU

CARBURETTORS

Des Hammill

VELOCE PUBLISHING PLC
PUBLISHERS OF FINE AUTOMOTIVE BOOKS

Other Veloce publications -

Colour Family Album Series
Alfa Romeo by Andrea & David Sparrow
Bubblecars & Microcars by Andrea & David Sparrow
Bubblecars & Microcars, More by Andrea & David Sparrow
Citroen 2CV by Andrea & David Sparrow
Citroen DS by Andrea & David Sparrow
Fiat & Abarth 500 & 600 by Andrea & David Sparrow
Lambretta by Andrea & David Sparrow
Motor Scooters by Andrea & David Sparrow
Porsche by Andrea & David Sparrow
Triumph Sportscars by Andrea & David Sparrow
Vespa by Andrea & David Sparrow
VW Beetle by Andrea & David Sparrow
VW Beetle/Bug, Custom by Andrea & David Sparrow
VW Bus, Camper, Van & Pick-up by Andrea & David Sparrow

SpeedPro Series
4-Cylinder Engine Short Block for High Performance, How to Blueprint & Build a by Des Hammill
Alfa Romeo Twin Cam Engines, How to Power Tune by Jim Kartalamakis
BMC/BL/Rover 998cc A-Series Engines, How to Power Tune by Des Hammill
BMC/BL/Rover 1275cc A-Series Engines, How to Power Tune by Des Hammill
Camshafts for Maximum Power, How to Choose & Time by Des Hammill
Cylinder Heads, How to Build, Modify & Power Tune (2nd edition) by Peter Burgess
Distributor-type Ignition Systems, How to Build & Power Tune by Des Hammill
Ford SOHC 'Pinto' & Sierra Cosworth DOHC Engines, How to Power Tune by Des Hammill
Harley-Davidson 1340 Evolution Engines, How to Build & Power Tune by Des Hammill
Jaguar XK Engines, How to Power Tune by Des Hammill
MGB 4-Cylinder Engines, How to Power Tune by Peter Burgess
MGB, MGC, MGB V8, How to Improve by Roger Williams
MGB V8 Power, How to Give your (Updated & Revised Edition) by Roger Williams
MG Midget & Austin-Healey Sprite for Road & Track, How to Power Tune by Daniel Stapleton
Road Car, How to Plan & Build a Fast by Daniel Stapleton
Sportscar/Kitcar Suspension and Brakes, How to Build and Modify by Des Hammill
SU Carburettors for High Performance, How to Build and Modify (Revised Edition) by Des Hammill
Weber DCOE & Dellorto DHLA Carburetors, How to Build & Power Tune (2nd edition) by Des Hammill
VW Beetle, How to Modify Suspension, Brakes & CHassis for High Performance by James Hale
V8 Engine Short Block for High Performance, How to Blueprint & Build by Des Hammill

General
Alfa Romeo Berlinas (Saloons/Sedans) by John Tipler
Alfa Romeo Giulia Coupe GT & GTA by John Tipler
Automotive Mascots: A Collectors Guide to British Marque, Corporate & Accessory Mascots by David Kay & Lynda Springate
Bentley Continental Corniche & Azure 1951-1998 by Martin Bennett
British Cars, The Complete Catalogue of 1895-1975 by Culshaw & Horrobin
British Trailer Caravans & their Manufacturers 1919-1959 by Andrew Jenkinson
British Trailer Caravans & their Manufacturers from 1960 by Andrew Jenkinson
Bugatti Type 40 by Barrie Price & Jean Louis Arbey
Bugatti 46/50, Updated & Revised Edition by Barrie Price
Chrysler 300 - America's Most Powerful Car by Robert Ackerson
Cobra - The Real Thing! by Trevor Legate
Cortina - Ford's Best Seller by Graham Robson
Daimler SP250 'Dart' by Brian Long
Datsun/Nissan 280ZX & 300ZX by Brian Long
Datsun Z - From Fairlady to 280Z by Brian Long
Dune Buggy Handbook, The by James Hale
Fiat & Abarth 124 Spider & Coupe by John Tipler
Fiat & Abarth 500 & 600 (revised edition) by Malcolm Bobbitt
Ford F100/F150 Pick-up by Robert Ackerson
Jaguar XJ-S by Brian Long
Jim Redman - Six Times World Motorcycle Champion by Jim Redman
Grey Guide, The by Dave Thornton
Lea-Francis Story, The by Barrie Price
Lexus by Brian Long
Lola - The Illustrated History (1957-1977) by John Starkey
Lola T70 - The Racing History & Individual Chassis Record New Edition by John Starkey
Lotus 49, The Story of a Legend by Michael Oliver
Mazda MX5/Miata 1.6 Enthusiast's Workshop Manual by Rod Grainger & Pete Shoemark *Mazda MX5/Miata 1.8 Enthusiast's Workshop Manual* by Rod Grainger & Pete Shoemark
Mazda MX5 - Renaissance Sportscar by Brian Long
MGA by John Price Williams
Motor Museums - of the British Isles and Republic of Ireland by David Burke & Tom Price
Mini Cooper - The Real Thing! by John Tipler
Porsche 356 by Brian Long
Porsche 914 & 914-6 by Brian Long
Porsche 924 by Brian Long
Rolls-Royce Silver Shadow & Bentley T-Series Updated & Revised Edition by Malcolm Bobbitt
Rolls-Royce Silver Spirit, Silver Spur and Bentley Mulsanne by Malcolm Bobbitt
Rolls-Royce Silver Wraith, Dawn & Cloud/Bentley MkVI, R & S-Series by Martyn Nutland
Singer Story: Cars, Commercial Vehicles, Bicycles & Motorcycles by Kevin Atkinson
Tales of Triumph Motorcycles by Hughie Hancox
Taxi! The Story of the 'London' Taxicab by Malcolm Bobbitt
Triumph Tiger Cub Bible by Mike Estall
Veloce Guide to the Top 100 Used Touring Caravans by Andrew Jenkinson
Velocette Motorcycles - MSS to Thruxton by Rod Burris
VW Bus, Camper, Van, Pickup, Wagon (Type 2) by Malcolm Bobbitt
Volkswagens of the World by Simon Glen
Works Rally Mechanic, Tales of the BMC/BL Works Rally Department 1955-1979 by Brian Moylan

First published 2000, reprinted 2001. Veloce Publishing Plc., 33, Trinity Street, Dorchester DT1 1TT, England. Fax: 01305 268864/e-mail: veloce@veloce.co.uk/ website: www.veloce.co.uk

ISBN 1-901295-14-1/UPC 36847-00114-8

Readers with ideas for automotive books, or books on other transport or related hobby subjects, are invited to write to Veloce Publishing at the above address.

British Library Cataloguing in Publication Data -
A catalogue record for this book is available from the British Library.

Typesetting, design and page make-up all by Veloce on AppleMac.
Printed and bound in the UK.

Contents

Veloce *SpeedPro* books -

ISBN 1 901295 45 1

ISBN 1 874105 76 6

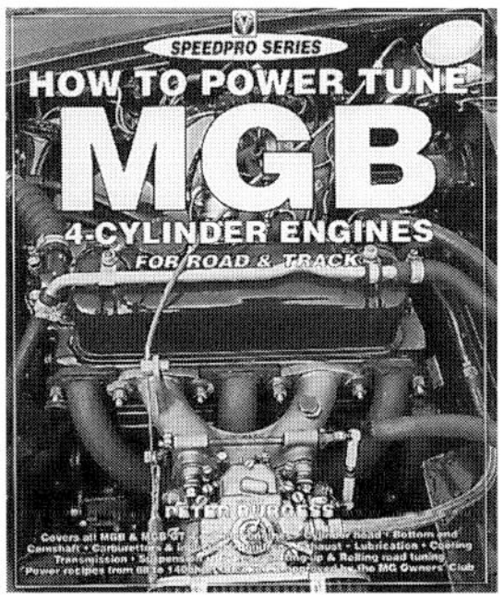

ISBN 1 874105 61 8

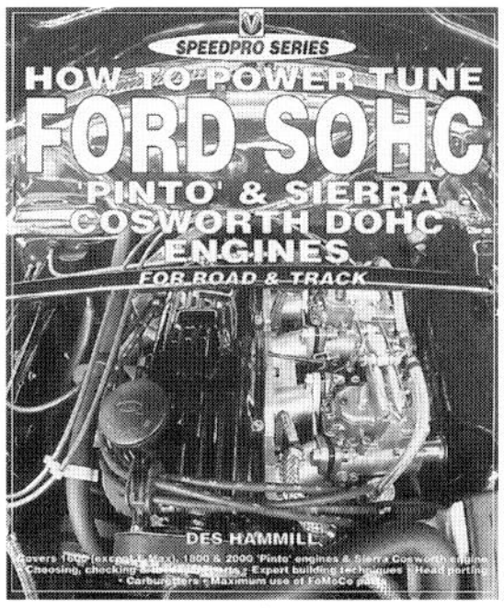

ISBN 1 874105 81 2

ISBN 1 901295 73 7

ISBN 1 901295 64 8

ISBN 1 901295 62 1

ISBN 1 874105 70 7

ISBN 1 874105 44 8

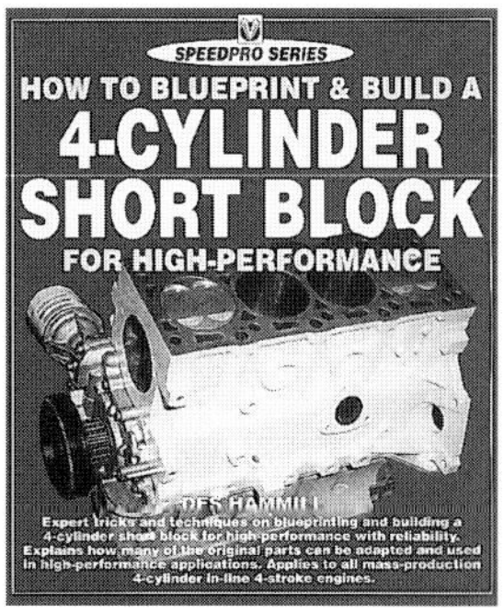

ISBN 1 874105 85 5

ISBN 1 874105 88 X

ISBN 1 901295 26 5

ISBN 1 901295 07 9

ISBN 1 901295 19 2

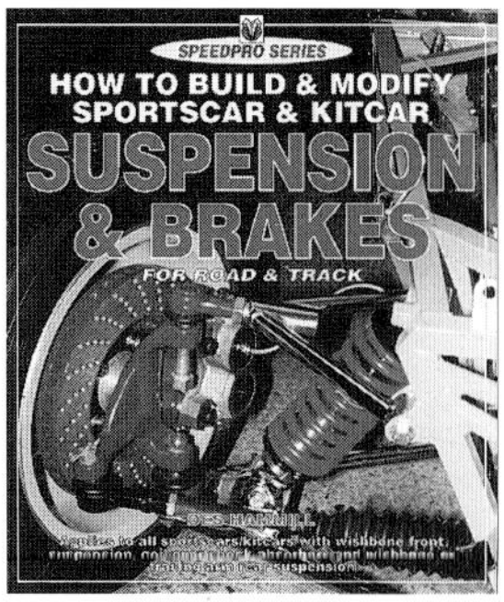

ISBN 1 901295 08 7

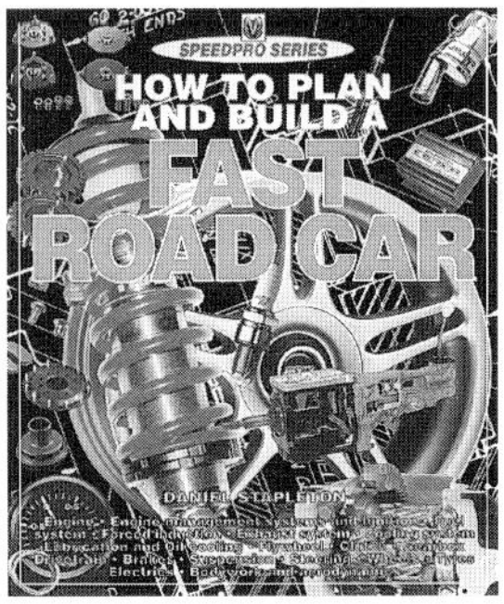

ISBN 1 874105 60 X

ISBN 1 901295 76 1

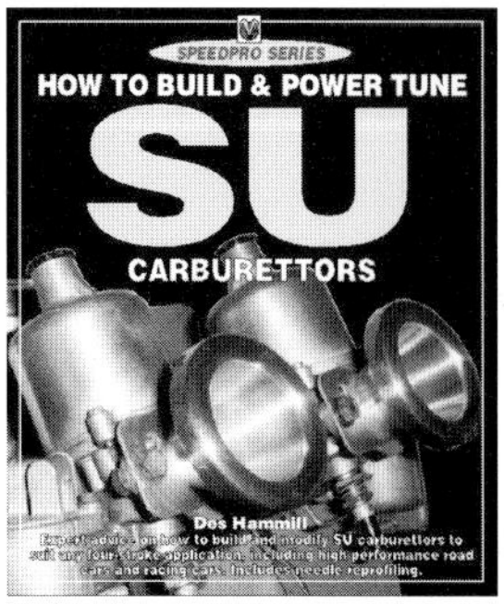

ISBN 1 901295 14 1

ISBN 1 901295 80 X

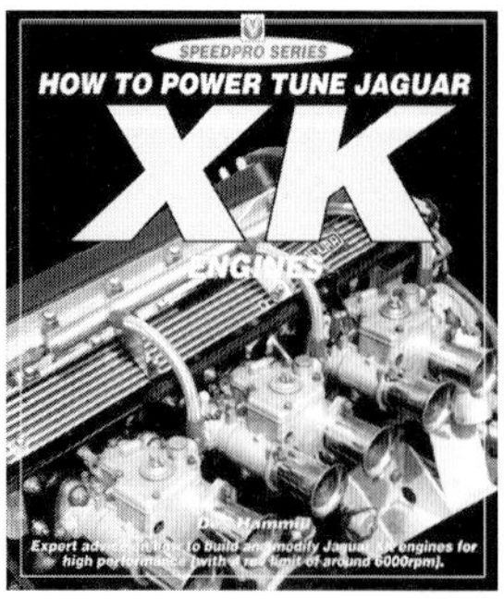

ISBN 1 901295 63 X

- more on the way!

Introduction

This book is for those people who wish to tune SU carburettors themselves, irrespective of how many carburettors there are on the engine or what type of engine it is you are dealing with. This book explains how to set up carburettors and choose needles *for any application,* and how to test the engine to make sure that the induction system is tuned for optimum performance.

The SU is a well made carburettor which can give excellent all round performance *and* excellent economy. SU carburettor sizes have varied over time but for the last thirty or more years the sizes have predominantly been 1 1/4in, 1 1/2in, 1 3/4in and 2in, with a few other less well known sizes having been made such as 1 1/8in and, more recently, 1 7/8in, 38mm and 44mm. This is a fair size range and while, singularly, an SU may not be able to supply the amount of air/fuel required for a given application, twin, treble or quad set-ups certainly can.

The majority of engines that these carburettors have been used on were standard mass production units with, usually, only one SU fitted, though some high performance cars such as the E-Type Jaguar (triple SUs) and the Mini-Cooper (twin SUs) used them too. A lot of four cylinder in-line engined sports saloon cars and sportscars have been factory fitted with twin SUs over the years (especially those from BMC/BLMC). With engines fitted with twin carburettors, the usual problem is one of butterfly synchronisation. The solution to this problem is to use a meter to check the airflows and reset the carburettors if necessary. The standard jetting is always correct for a standard specification engine, but even a small modification such as a modified cylinder head can mean that the standard needles are no longer correct for the application. It often pays to go through the suggested tuning sequence in chapter seven of this book just to be sure that you are actually getting the most out of your particular engine.

Carburettor components get replaced over the years and it is not impossible that vital parts (needles and piston springs, for example) have been changed at some time and that a given carburettor is now nowhere near standard. The first thing anyone should do is check to see that the standard recommended jetting is, in fact, in place to start with. It's amazing how many older cars no longer have the standard recommended jetting in their carburettors!

Innovative people and businesses have fitted these carburettors to all manner of engines over the years using custom fabricated or specialist aftermarket cast aluminium inlet manifolds. The British Rover V8, for instance, can be fitted with four SU carburettors (two are the normal factory fitting) using a steel fabricated inlet manifold. The four SU system works well with this engine, although care must be taken when synchronising the four butterflies at idle and off idle. Overall the system is very good with fuel distribution being significantly improved over the standard inlet manifold. The system is expensive new but second-hand the price is much more attractive. This particular induction system serves to illustrate what is possible with these carburettors.

With these carburettors the high performance capability is certainly there, but the high performance image isn't (or, at least, for most people it isn't), especially when it comes to comparing SUs to sidedraught Webers and Dellortos. To most eyes a pair of SUs on a four cylinder in-line engine do not look as good as a pair of twin choke sidedraughts, yet the difference in performance is often much less in real terms than people imagine, especially in terms of overall engine efficiency. The SU can be a serious high-performance carburettor and at a fraction of the cost of Webers or Dellortos. Plenty of people still want to use SUs on their engines and want the carburettors to be set up correctly.

It's certainly wrong to write-off the SU as being totally inferior to sidedraught Webers or Dellortos (or any other carburettor for that matter) on the basis of being capable of delivering the correct air/fuel ratio and volume of air/fuel for a high performance application. SU carburettors can match any carburettor on this basis, but what they can't do is match sidedraughts for engine acceleration and overall engine response (accelerator pump and instantaneous air/fuel supply) - unless the SU is modified. The SU can be modified quite easily to improve engine acceleration.

Ideally a single SU per cylinder would be fitted, but this is seldom ever done. The Japanese motorcycle industry almost universally uses constant depression carburettors (similar to the SU) and one carburettor per cylinder and there are certainly not too many complaints about Japanese motorcycle engine performance! The situation with the Japanese motorcycle engines is, of course, that the carburettors are precisely sized for use on the particular engine and the throttle linkage is perfect in all respects.

On the road the SU is always as good as any other carburettor. It will almost always deliver superior fuel economy (no accelerator pumps pumping in fuel for a start) to any other carburettor available for the same engine. The SU certainly excels here and really has few rivals for delivering very acceptable power at a most economical rate of fuel consumption - provided the needle is the correct one for the application.

It's fair to say that when an engine is fitted with a modified cylinder head with enlarged inlet ports, the fuel economy will suffer no matter how gently the car is driven. Power will almost always increase - especially as the revs rise - after fitting a modified cylinder head or heads and related high performance components. Increasing the inlet port size is the main factor in the increase in fuel consumption because, at all times, more air/fuel can be admitted to the engine through increased cylinder head efficiency. Even at idle more air/fuel mixture will be admitted.

High performance camshafts can also adversely affect fuel economy, especially when used with a modified cylinder head. The situation is not quite the same if only a camshaft change has been made: the reason for this is that, provided the camshaft change was a sensible one, the inlet ports of an unmodified cylinder head are effectively upstream and are still a limiting factor. If the camshaft choice is completely unrealistic nothing can save the engine from being a 'pig,' and while the engine will consume a huge amount of fuel, it won't go well. Engines must be modified as a package, with no one thing being the 'magic' answer.

Many lack of power problems associated with SUs can be traced to the inlet manifold and, more specifically, to the cross-sectional area of the inlet manifold runners. The solution to this problem can be as simple as porting the inlet manifold runners to a suitable size and, perhaps, improving the contours of the internal passageways of the inlet manifold (smooth curves/turns). The same inlet manifold may work fine on the standard engine, but fall short when the engine is modified because insufficient air can be moved through it to match the new capability of the modified engine. The result is a lack of power, but this is not the fault of the carburettors or the engine *per se*.

Many engines have no expense spared on the cylinder head modifications, yet retain near standard internal size inlet manifolding! Further to this, many of the high performance manifolds available have insufficient runner cross-sectional area and, although they may look right from the outside, are often inadequate on the inside, yet it is the inside that really matters.

Take, for example, a twin carburettor four cylinder in-line engine fitted with a cylinder head with round inlet ports. Firstly, the diameter of the inlet manifold where the carburettor/s bolt on to it needs to be the same size as the carburettor throttle bore diameter. Secondly, as a rough guide,

A standard-type cast iron inlet manifold which has been ported out as much as possible.

each individual runner must not be smaller in cross-sectional area than the individual port diameters of the cylinder head to inlet manifold matching face. This is a little bit of an over-simplification as it assumes that the port sizes of the cylinder head are correct in the first place. It is however actually quite difficult to over-port an inlet manifold to the detriment of power production. The majority of cast aluminium inlet manifolds have reasonable wall thickness, but there is always a limit to how much material can be taken away, although such manifolds can be Tig welded quite easily if breakthrough occurs. The same applies to cast iron inlet manifolds though they can be brazed.

If the individual inlet manifold runners are opened out considerably, the amount of material left can become a bit thin. The solution to this problem is to weld gussets onto the manifold to strengthen it, otherwise it is possible to port out an inlet manifold so much that it can crack through a lack of material strength and the carburettor/s will literally fall off the engine!

Note that if the inlet manifold internal sizing is not matched to suit the carburettors (and that means everywhere within the inlet manifold) only a minor increase in power will be achievable.

SUs are inexpensive second-hand and are readily available from scrapyards/car breakers everywhere. It takes a fair few miles/kilometres to wear out an SU, but the telltale sign of wear for any SU is spindle movement (which indicates a lack of correct running fit between the throttle spindle and the carburettor body). The wear can be confined to the brass spindle or be a combination of carburettor body wear and spindle wear. With the amount of second-hand SUs available it's almost always never necessary to completely rebuild a carburettor (to replace the spindle and re-bush the carburettor body to restore the original running fit). In the overall scheme of things, if you have an SU carburettor which is of a common type and it has a worn spindle/body, go and get another good replacement carburettor from a scrapyard/breaker. If the carburettors you have are not so common, consider rebuilding them using all new parts or, alternatively, buy new old-type carburettors. This book assumes that the carburettors being used are in perfect working order with no excessive wear.

If you have freedom of choice, the later design of SU is definitely better and is the model to go for. This is the HIF model (HIF stands for Horizontal, Integral, Floatchamber). Having the fuel bowl integral with the carburettor body, instead of hanging off the side, and adjustment of the seat by a side-positioned screw, rather than a conglomeration of nuts, springs and seals underneath the carburettor, are good evolutionary improvements. HIF SUs for example, have a side load on the needle and, as a consequence, on the bore walls of the main jet. This means that the main jet wears and loses its original sizing, becoming oval in the process (needles do not wear at anything like the same rate).HIF needles are 'spring-loaded,' not 'fixed.' These HIF carburettors are neat and tidy. The other point, of course, is that there is much less chance of fuel leaks with the new design. Also the dashpot is held in place by a circlip to prevent it falling out when the top of the carburettor is removed. All in all, SU have sorted out the problems with these carburettors

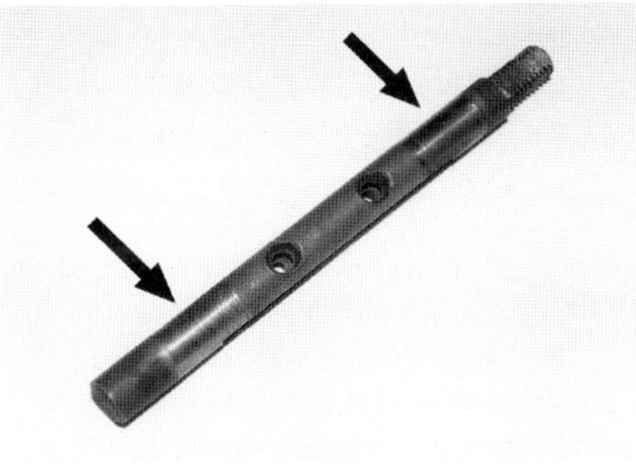

Well worn (0.007in/0.17mm) SU spindle: replace it.

HIF baseplate which has an O-ring seal.

Main jet on an HIF carburettor is raised or lowered by this screw.

HIF: Dashpot spindle uses circlip retention.

to the point that the HIF is virtually faultless. The earlier carburettors are just as good in performance terms (when in good condition) as the operating principles remain the same.

No claim is made that SUs produce more power than other carburettion systems, but there have been many instances where SU-equipped engines have performed as well as Weber/Dellorto sidedraught-equipped engines.

SU carburettors suffer from the same problems, such as throttle spindle wear and carburettor body wear, as other volume production carburettors. The SU does not wear out quickly and it is only after it has covered a huge distance that it shows signs of significant wear. Carburettors with worn spindles and bodies can cause fluctuating idle speeds, or an inconsistent idle which can be annoying rather than detrimental. All SUs are accurately machined when made and can be considered to be precision instruments. Any wear will manifest itself as a power loss in the final analysis. Things like spindle wear, throttle linkage wear, main jet and needle wear will all play their part in reducing the power potential of the engine. For high performance applications, use only SUs that are in as-new condition.

Synchronisation can be a problem with multiple carburettors,

Latest HIF44 SU carburettor.

but most factory-made linkages are substantial enough to withstand considerable abuse and not allow the carburettors to become 'out of synch.' The usual problem is where the carburettor spindles are seriously worn (0.005in/0.125mm to 0.010in/0.25mm, or more). What happens then is that the spindles seldom settle back in the same place twice at idle. So, after carefully adjusting the idle speed of both carburettors to, say, 900rpm, it is found that after the first 'blip' or two of the throttle the idle is too high or too low (anything from 1300rpm to 500rpm). This problem only affects idle and will mean the spindle/s have to be replaced at least, and if the carburettor body spindle bores are worn too the replacement spindle/s will only improve the situation to a certain degree. The carburettor body can be rebushed to restore the original spindle bore size, but this is seldom done because of the easy availability of inexpensive second-hand carburettors.

The off idle adjustment synchronisation is only concerned with making sure that both butterflies are set identically on the basis of airflow. This means that at any engine speed apart from idle all carburettors are showing the same amount of airflow. This is measured with a meter (of which there are several types). The days of using a tube and listening to the hiss of the carburettors are long gone (thank goodness): don't waste your time with such methods.

Inexpensive meters are available (such as Gunson's Carbalancer) through to the more expensive Motometer. The Gunson's Carbalancer is dead accurate and is ideal for synchronising all SUs. This product is inexpensive and available worldwide. Product information is available from Gunson's head office in England - Gunson Limited, Coppen Road, Dagenham, Essex, England RM8 1NU (Phone: 0181 984 8855/ Fax: 0181 984 8866). In the USA from the Actron Manufacturing Company - Actron Manufacturing Company, 9999 Walford Avenue, Cleveland, Ohio 44102 (Phone : (216) 651 9200/Fax: (216) 651 2388).

If the carburettors are not synchronised the engine will not

Motometer airflow meter.

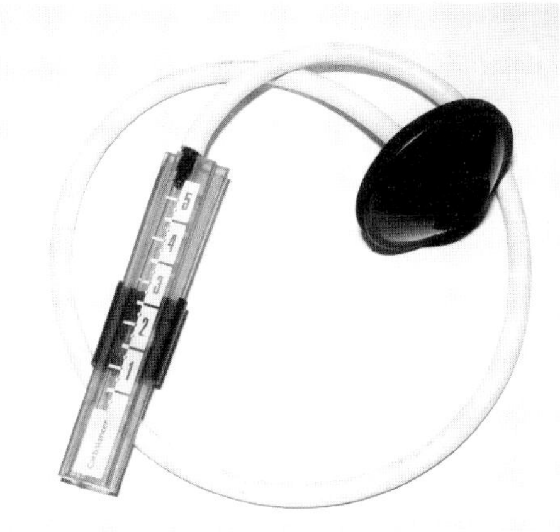

Gunson Carbalancer airflow meter.

develop full power or go anything like as well as it should. The reason is that the amount of air being admitted to the engine via each carburettor is not equal. On a four cylinder in-line engine, for example, the front two cylinders could be receiving 10% more air than the rear two cylinders and the engine is essentially 'fighting itself' with two cylinders trying to run at one speed and two trying to run at another speed. The synchronisation of SU carburettors needs to be checked on a regular basis: every six months for road cars. Certainly any engine using SUs fitted to a racing car should have the off idle synchronisation checked before each race day. Taking the time and trouble to get the off idle synchronisation dead right is paramount. This process is always much easier when the carburettor spindle bores and the spindles themselves are in excellent condition. If trouble is experienced synchronising SU carburettors always check the spindles and bodies first, then the linkages. Renew whatever is wrong.

A very common complaint with multiple SUs is that they often seem to need frequent adjustment. Experience has shown that this is only true when parts, such as the spindles/carburettor body bores and linkages, are worn. A robust throttle linkage with no slop in it is essential.

Repair kits and parts for all types of SU are readily available worldwide. Genuine SU parts are still being made by a British based company called Burlen Fuel Systems Ltd, an SU licensed company dedicated to supplying all SU parts. Burlen Fuel Systems will supply parts to anyone anywhere in the world and will supply you direct if there is no agent available. All parts come packaged as genuine SU parts and they do make *every* conceivable part. To all intents and purposes Burlen Fuel Systems can be considered to be SU. Their address is - Burlen Fuel Systems Ltd, Spitfire House, Castle Road, Salisbury, Wiltshire, England SP1 3SA (Phone: 01722 412500/Fax: 01722 334221).

There are more than 800 needle types available for SUs. There is a needle for nearly every application and basically more than one needle that will suit the same application. The differences in the needles are often very slight. The selection of the needle is the critical factor in the tuning of an SU as, if the needle is not correct for the application, the engine will never run well. Mainly factory made needles are mentioned in this book. The manufacturing of needles is a highly precise business and, although needles can be polished to reduce their diameters, most people do not do this and instead select needles from the *SU Carburettor Needle Profile Charts* book. However, there is a complete chapter in this book about how to successfully alter standard SU needles to suit any requirement.

Selecting needles is the part of tuning SUs that can become a bit frustrating, and is probably the reason why these carburettors have been largely passed over for high performance applications. To have to go and get new needles when the ones just fitted don't work as expected can be a bit annoying. That said, the tuning of other types of carburettors (sidedraught Webers/Dellortos for instance) is certainly not inexpensive and the wrong choice of parts most certainly gets expensive in a hurry: there are plenty of engines around which are not running correctly simply because the owner of the car won't spend the necessary money to buy the right carburettor bits. One factor in the SU's favour is the fact that, primarily, only one component is changed to alter the fuelling characteristics: the needle (and, to a lesser extent, the piston spring).

Effectively, selection of a new needle is based on the size of the needle that has just come out of the carburettor, and where in the rpm range the mixture is wrong. There is no reason whatsoever why the right needle for the engine and application cannot be found as there are only so many combinations (within reason) that are going to work. The richness/leanness factor of the needle is controlled by its diameter and this aspect of a needle can be successfully determined by using the *SU Carburettor Needle Profile Charts* book to find a richer or leaner needle. If standard SU needles are going to be used, measuring of the actual needle is unnecessary as all SU needles are letter and/or number coded, but efficient use of the relevant SU chart **is** necessary. The modifying of existing needles is often required if the tuning work is to be carried out quickly.

While the range of needles as made by SU is extensive, it does not necessarily cover all of the requirements of a modified engine. In this situation the only solution is to actually 'make' a needle to suit the application. This process comprises

The three books available from Burlen Fuel Systems (see text).

nothing more than selectively polishing an existing needle in certain positions. This is not all that difficult to do, but does require accurate measurement of the needle diameter (at the right distance down from the shoulder of the needle) using a micrometer coupled with some dexterity. All things considered, modification of needles to suit any application is not a daunting task.

There is a sequence to successfully tuning SUs. This sequence is not all that complicated but it can be quite time-consuming, with some of that time being used to procure alternative needles. While the tuning of SU carburettors can be quite frustrating, it can also be very satisfying to do it yourself - even if it does take a bit longer than you might have hoped.

There are many examples of high performance engines using SU carburettors very successfully and with no aspect of the engine performance lacking at all. These carburettors can be used with absolute confidence.

PREPARATION

No engine can be tuned correctly if it has a mechanical problem (worn, maladjusted or broken componentry) or if the electrical system has a fault.

Before embarking on tuning carburettors, replace any well used engine component and check the integrity of the electrical system. This means new sparkplugs, sparkplug leads, distributor cap, rotor, points, condenser and coil, and checking the condition of all associated wiring. If the distributor is electronic the module should be checked (using a module tester) to see that it is working at 100%. After all of this, the amount of voltage present at the actual sparkplug should also be checked.

There are some very simple, inexpensive gadgets available that can check the overall efficiency of the ignition system. The Gunson Flashtest, for example, can check the amount of spark voltage going to the sparkplug. This gadget fits into the end of the plug lead and then on to the sparkplug cap. This means that if the spark is within the range on the scale which indicates a good spark everything is correct with the electrical system right up to the top of the sparkplug. It does not, of course, tell you the condition of the sparkplug itself or how well it is firing. **Caution!** - If the Gunson device is used to test an electronic ignition system, it must only be used within the limits of the scale (24KV). **Do not** move the arm out of the stop to see just what gap the spark will actually jump. The result of this will be a blown module. 24KV is quite enough voltage to have on the scale as an indication of the integrity of the spark when using this particular piece of testing equipment.

The minimum amount of voltage that should ever be considered appropriate is approximately 15KV, with up to 24KV being possible with the better systems. 15+KV is enough to successfully fire most high compression engines. If your engine has KV readings at the sparkplugs of less than 11KV do not expect to get good engine performance. With the Gunson Flashtest unit the green (good) reading starts at 15KV.

Engines which have a low sparkplug voltage run right through to maximum rpm, but they are sluggish, that's the difference. Often plenty of money is spent on every other aspect of an engine, yet this vital component of engine performance is not even checked. The higher the voltage going to the sparkplugs (within reason, of course) the better the engine will go. This voltage thing cannot be stressed enough: too many engines do not have good electrical systems.

Veloce Publishing has further SpeedPro title by Des Hammill called *How To Build & Power Tune Distributor-type Ignition Systems,* which will prove of benefit to anyone who has this sort of ignition system fitted to their engine.

Any engine which has retarded ignition timing will never develop maximum power or pull well through the rpm range. Retarded ignition timing is a major cause of power loss when it comes to modified engines: there is virtually no engine around today that runs properly on anything less than about 30 degrees of total advance (over 3500rpm) under full throttle conditions, with many engines needing up to 38 degrees of total advance and an odd few even more. Each engine is different in its requirements so it is not possible to give hard and fast rules. Altering the ignition advance can be as simple as moving the distributor on some engines but, more often than not, it requires altering the distributor internally. Engines with retarded ignition are poor performing engines.

Using this book & essential information

USING THIS BOOK

Throughout this book the text assumes that you, or your contractor, will have a workshop manual specific to your carburettor to follow for complete detail on dismantling, reassembly, adjustment procedure, clearances, etc. This book's default is the standard manufacturer's specification for your carburettor model so, if a procedure is not described, a measurement not given, you can assume that the standard manufacturer's procedure or specification for your carburettor needs to be used.

You'll find it helpful to read the whole book before you start work or give instructions to your contractor. This is because a modification or change in specification in one area may cause the need for changes in other areas. Get the whole picture so that you can finalize specification and component requirements as far as is possible before any work begins.

For those wishing for information on high-performance short block building principles, ignition systems, cylinder head work and camshafts, the following Veloce titles are recommended further reading. *How To Blueprint & Build a 4-Cylinder Short Block For High Performance, How To Build & Power Tune Distributor-type Ignition Systems, How To Build, Modify & Power Tune Cylinder Heads, How to Choose Camshafts & Time them for Maximum Power.*

ESSENTIAL INFORMATION

This book contains information on practical procedures; however, this information is intended only for those with the qualifications, experience, tools and facilities to carry out the work in safety and with appropriately high levels of skill. Whenever working on a car or component, remember that your personal safety must **ALWAYS** be your **FIRST** consideration. **The publisher, author, editors and retailer of this book cannot accept any responsibility for personal injury or mechanical damage which results from using this book, even if caused by errors or omissions in the information given. If this disclaimer is unacceptable to you, please return the pristine book to your retailer who will refund the purchase price.**

In the text of this book **"Warning!"** means that a procedure could cause personal injury and **"Caution!"** that there is danger of mechanical damage if appropriate care is not taken. However, be aware that we cannot foresee every possibility of danger in every circumstance.

Please note that changing component specification by modification is likely to void warranties and also to absolve manufacturers of any responsibility in the event of component failure and the consequences of such failure.

Increasing engine power will place additional stress on engine components and on the car's complete driveline: this may reduce service life and increase the frequency of breakdown. An increase in engine power, and therefore vehicle performance, will mean that your vehicle's braking and suspension systems will need to be kept in perfect condition and uprated as appropriate. It is also usually necessary to inform the vehicle's insurers of any changes to it's specification.

The importance of cleaning a component thoroughly before working on it cannot be overstressed. Always keep your working area and tools as clean as possible. Whatever specialist cleaning fluid or other chemicals you use, be sure to follow - completely - the manufacturer's instructions and if you are using petrol (gasoline) or paraffin (kerosene) to clean parts, take every precaution necessary to protect your body and to avoid all risk of fire.

Chapter 1

How an SU carburettor works

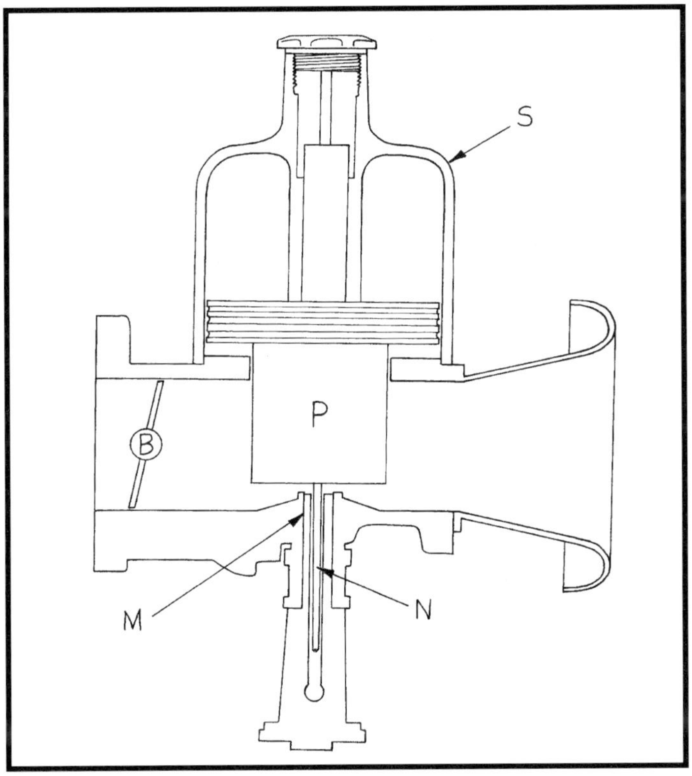

Butterfly "B", piston "P", suction chamber "S", main jet "M", needle "N."

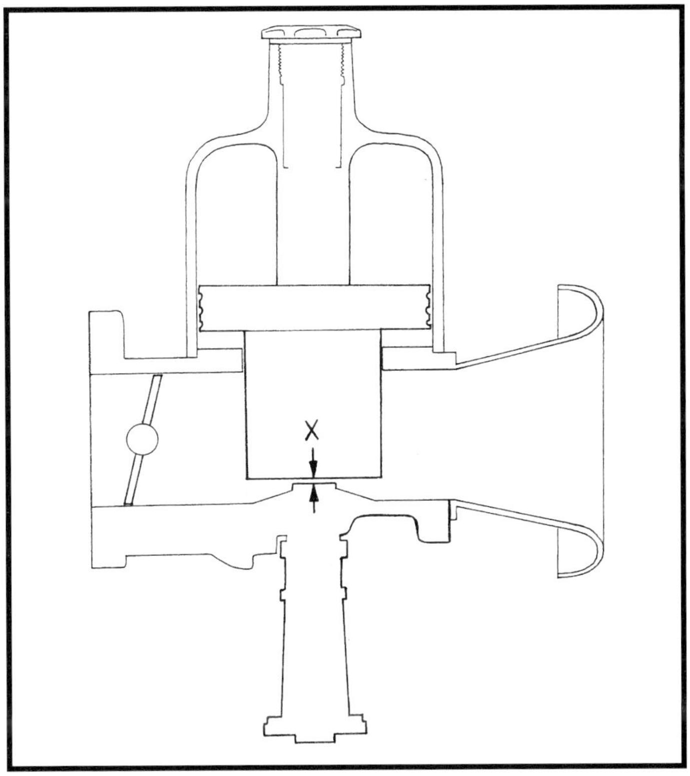

Piston base is never in full contact with the bridge of the carburettor. There is always a minimum 0.015-0.018 inch air gap "X" between the top of the bridge and the base surface of the piston.

Plastic stop (arrowed) in piston.

Piston on left has two slots and two drillings (arrowed), while piston on right has the less common single drilling in back face (arrowed).

These carburettors operate on the principle that when the butterfly is opened, more air can go into the engine, which is the same as most other carburettors. However, the air cannot go directly into the engine when the butterfly is opened because the SU has a piston assembly (which moves vertically) between the butterfly and the air entry point of the carburettor. This piston has to be lifted before any substantial amount of air can be admitted to the engine.

This means there are two distinct systems involved in admitting air into the engine. When the throttle is opened, the volume of air that could pass through the butterfly is determined by how much the butterfly is open. However, the air that passes through the butterfly has first to pass between the base of the piston and the bridge in the throttle bore of the carburettor. The piston position is the limiting factor to air flow to the engine.

There is no mechanical connection between the butterfly and the piston. The piston is lifted by engine generated suction, so it lifts off the bridge when there is sufficient suction and by an amount determined by the amount of suction acting upon it.

When an engine is at idle, the butterfly is closed and the piston is either resting on the bridge or riding slightly above the bridge. In fact, the piston can't actually rest directly on the top surface of the bridge because the plastic stop fitted to its base prevents it from doing so. Of course, at idle the butterfly is not quite shut and will be admitting a small amount of air to the engine. The plastic stop fitted into the piston base is proud of the surface by 0.015-0.018 inch. When the engine is switched off and no vacuum present there is always a minimum gap of 0.015-0.018 inch between the base surface of the piston and the top of the bridge. This means that a minimum amount of air can always pass between the base of the piston and the top of the bridge. So, when the engine is started, the piston does not have to be lifted by suction before an amount of air can be admitted to the engine.

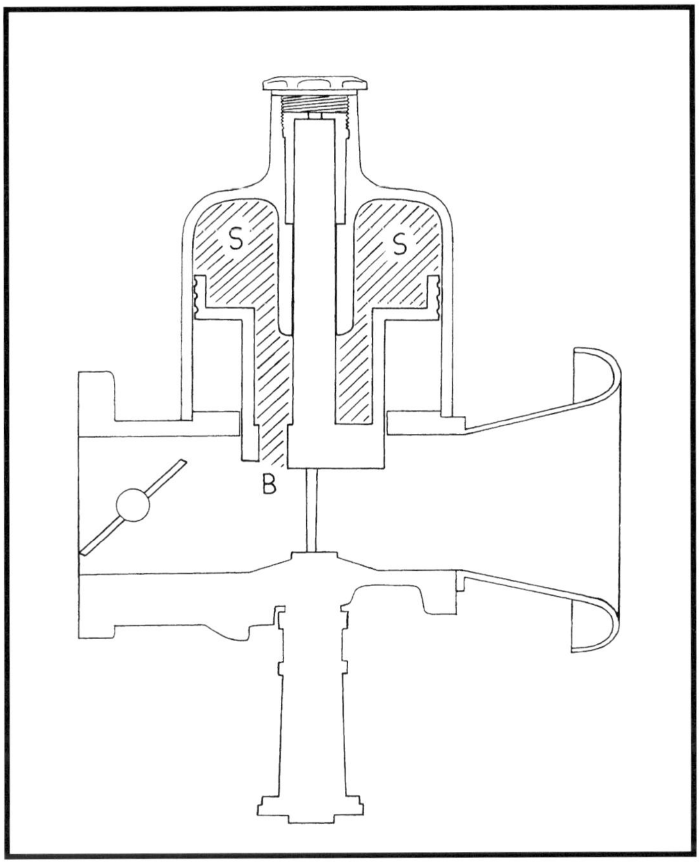

Engine suction (vacuum) "S" is applied to the chamber above the piston by two drillings "B."

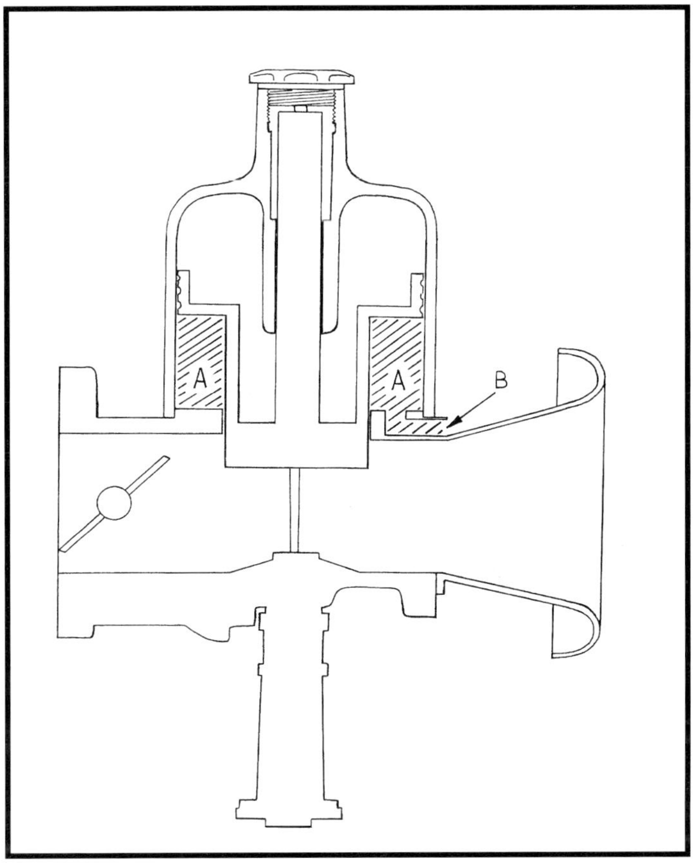

The air beneath the underside of the piston is vented to atmosphere at "B" (not to scale). Therefore the area beneath the piston is always at atmospheric pressure.

The suction (vacuum) link between the engine and the top of the piston in the suction chamber is via two drillings in the base or one drilling in the back edge of the piston. These drilled holes - which are on the downstream side of the piston (that's between the butterfly and the centre of the bridge) - allow engine generated suction to be transfered from this area of the carburettor to the top side of the piston in the suction chamber.

The underside of the piston in the suction chamber is vented to atmosphere and is subjected to near normal atmospheric pressure at all times. The suction chamber on the top of the piston will have varying amounts of suction working in it once the engine is running. So, when the engine is not running, there is atmospheric pressure acting on both sides of the piston in the suction chamber. In this situation the piston rests on the bridge of the carburettor body (actually the plastic stop).

The instant the engine is running and idling there will be less than atmospheric pressure on the top side of the piston and the piston will lift up by a certain amount. What lifts the piston is the atmospheric pressure acting on its underside versus the sub-atmospheric pressure (suction) above the piston.

To accelerate the engine from idle, the butterfly is opened further which allows more air to enter the engine which, in turn, creates more engine suction. Increasing suction lifts the piston higher and the piston lifts the needle, which also allows more fuel to be sucked through the main jet to mix with the increased air volume going into the engine.

The more engine suction there is, the higher the piston will lift. The limit for the piston travel is the physical limit imposed by the design of the carburettor. When the butterfly is fully open, and the engine has enough suction to fully lift the piston, the maximum mechanical airflow has been achieved.

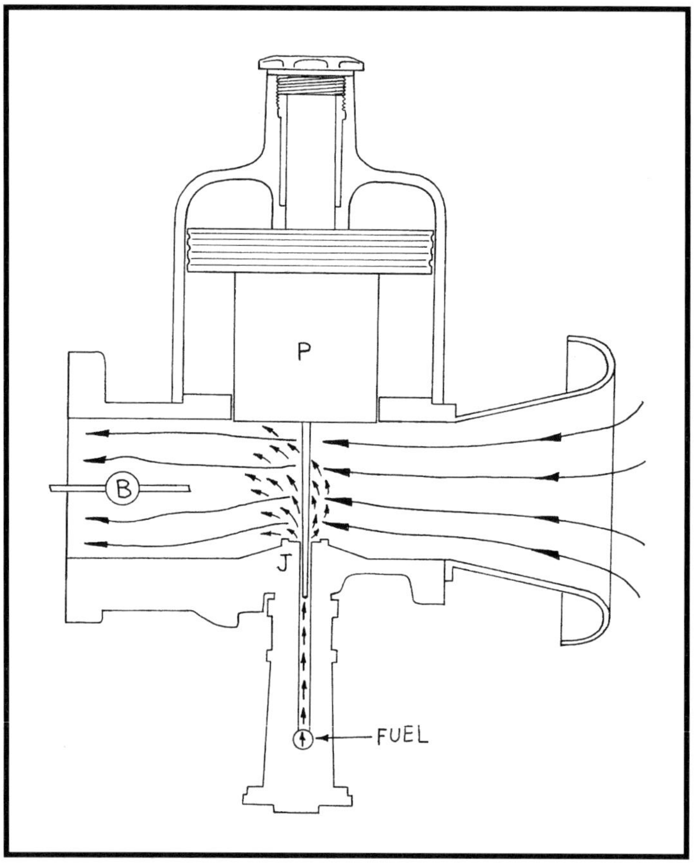

Full throttle is applied here (see butterfly "B"). The suction (vacuum) is high and piston "P" is fully raised so the needle is delivering maximum fuel supply via the jet "J."

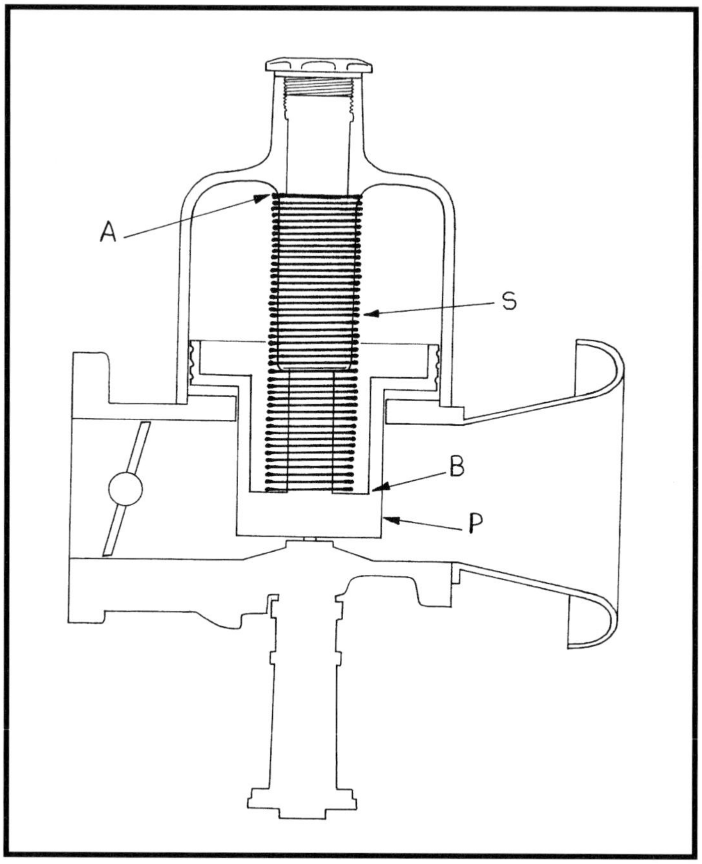

The piston spring "S" fits in the suction (vacuum) chamber between the top of the suction chamber "A" and the inner base of the piston "B". As the piston "P" rises it compresses the spring and the spring tension offers resistance to movement.

The amount of piston lift, for a given amount of engine suction and butterfly opening, is controlled by the resistance offered by the tension of the particular piston spring fitted to the carburettor. The stronger the piston spring tension, the lower the piston will ride in the carburettor throttle bore for a given amount of suction. Conversely, the weaker the piston spring is the higher the piston will ride in the carburettor throttle bore for a given amount of engine suction and butterfly opening.

The piston spring is, in effect, an economy device because, within sensible limits, the lower the piston rides for a given amount of butterfly opening, the leaner the mixture will be. This can be taken too far (by stronger springs), to the point that the fuel mixture is so lean the engine will not run correctly. For good engine performance and reasonable fuel economy, the lighter piston springs made by SU are recommended. All SUs equipped with piston springs and oil in the dashpot are relatively slow to accelerate in the lower to mid-rpm range.

The next part of this scenario is the amount of fuel being mixed with the air passing between the top of the bridge of the carburettor and the base of the piston, which has to be correct to sustain engine performance.

The working arrangement of the SU carburettor is that the needle is fitted centrally into the bottom of the piston and the main fuel discharge point is in the centre of the bridge of the carburettor body. The lift of the piston also controls the main jet fuel discharge area (the gap between the needle and the walls of the main jet) and, therefore, the amount of fuel that can pass up into the airstream. The bore of the piston and suction chamber and the main jet are all on the same centreline. Having the right needle profile is **vital** to the successful operation of an SU carburettor.

The vertical movement of the rising piston is damped so that there

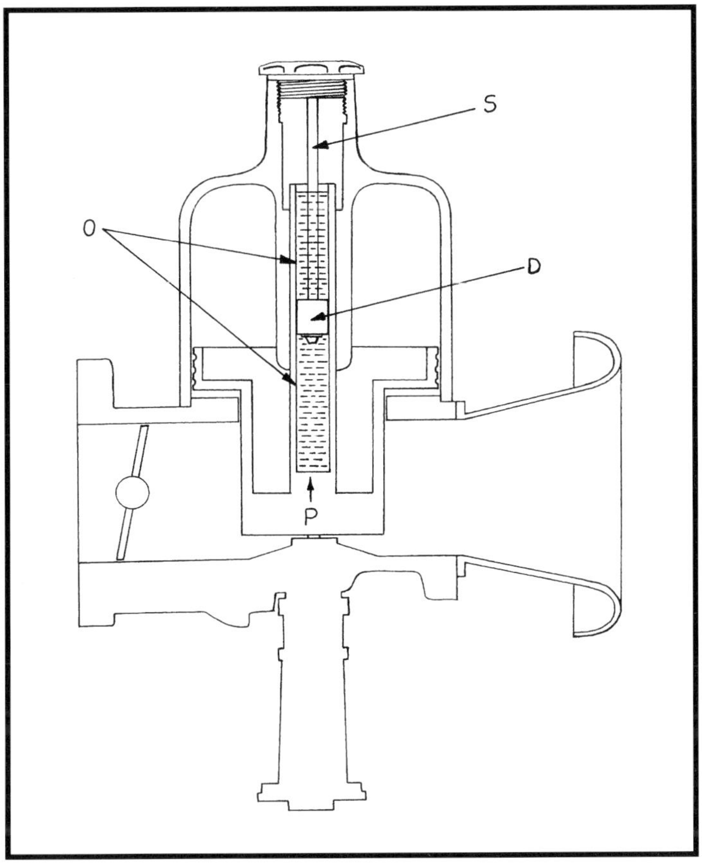

The position of dashpot damper "D" is fixed. There is oil above and below the damper. When the piston "P" rises and falls, the oil "O" passes through the gap created by the clearance between the drilling in the damper and the damper spindle "S."

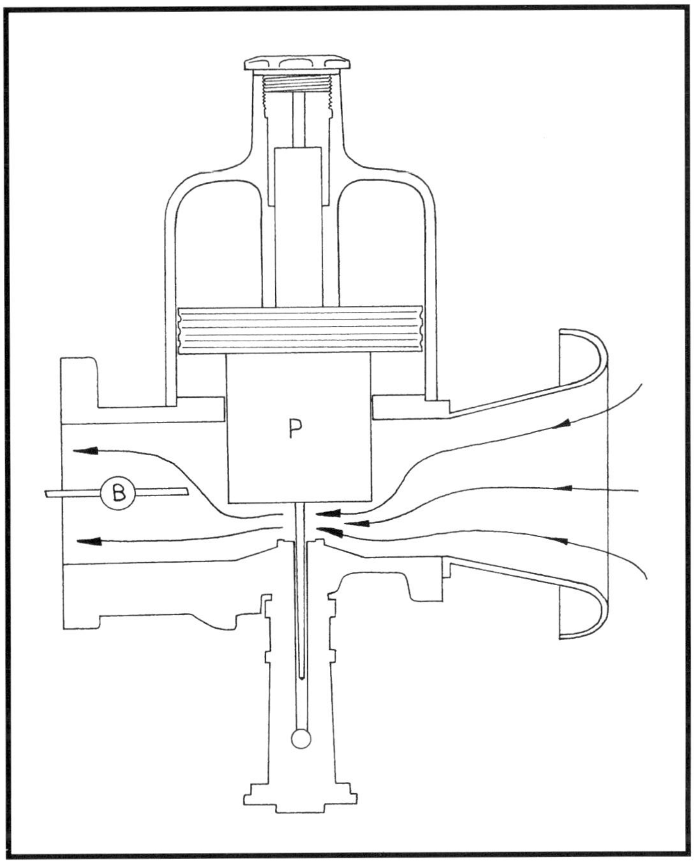

Carburettor at full throttle (see "B"). If the engine is under severe load it will not produce much suction (vacuum) and the piston "P" will not rise very much.

are no sudden rises which, in all normally tuned SUs, would result in engine hesitation (lean mixture) as the fuel mix will not adapt fast enough to the engine's requirements. The piston would rise virtually instantly in response to changing suction if it wasn't for the dashpot damping mechanism. The damping mechanism is comprised of a stationary damper immersed in oil. When the piston starts to rise the oil is forced around the damper, a process which slows the rise of the piston and the engine accelerates cleanly but slowly because of this system.

The amount of air being admitted to the engine by the butterfly, the amount of air passing between the bridge of the carburettor body and the base surface of the piston, and the amount of fuel being drawn out of the main jet have all to be matched. If the amount of fuel that can be drawn out of the main jet is insufficient (lean), the engine will falter and, even if the butterfly is opened fully, the piston will not lift because there is insufficient suction being generated by the engine. The main variable here is the needle. If the needle is wrong the engine will not run properly as there will be insufficient fuel being drawn into the engine to support combustion.

On a correctly tuned engine the air/fuel mixture is controlled by the height of the piston above the bridge in the throttle bore of the carburettor. The butterfly is used to initialise acceleration, but the piston will not rise unless there is sufficient suction acting on it. To illustrate this point further: if a car is going up hill in top gear with the butterfly fully open, it does not mean that the piston is as high as it can go. If such a scenario is taken a bit further, the engine could be labouring away with the rpm dropping off slowly and the driver does not change down when he or she should. The throttle (butterfly) can be wide open but the piston might well be up off the bridge by only a small amount (low suction). The speed of the car will

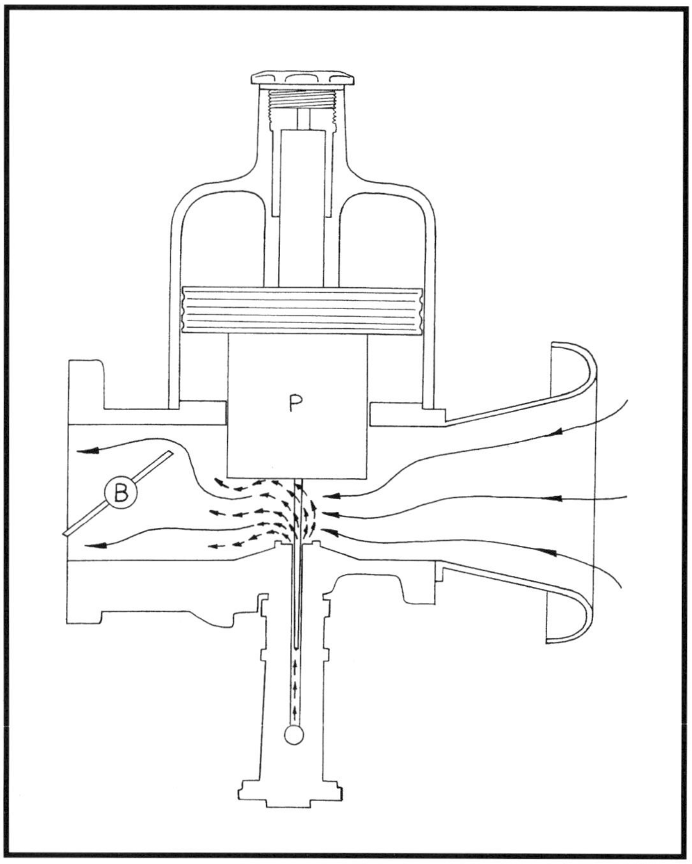

Carburettor at half throttle (see "B"). Fuel is drawn out of the main jet and atomised.

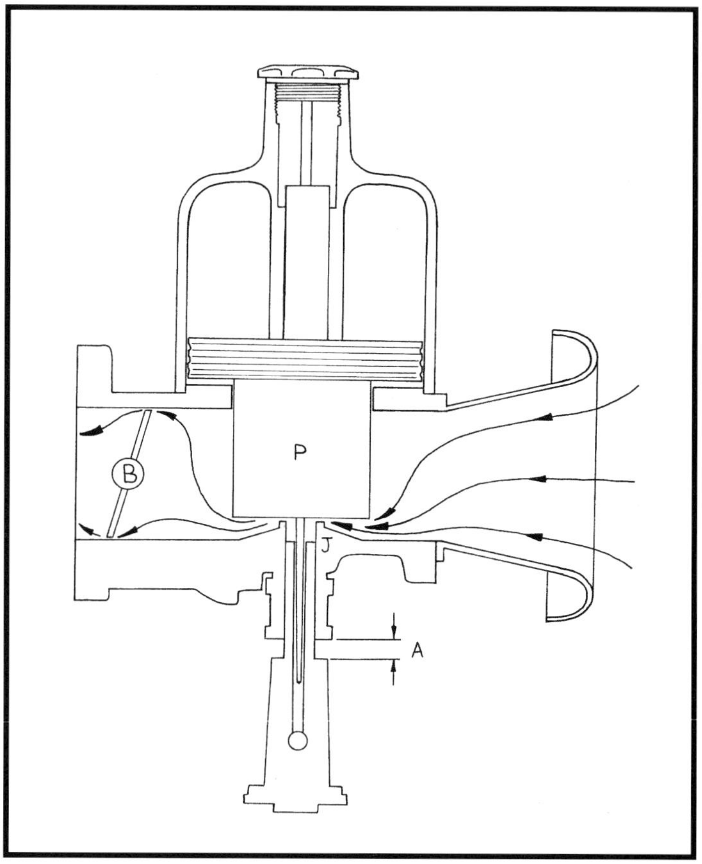

Carburettor at very small throttle opening (see "B"), with choke applied. The main jet has been lowered (0.25in) from the bridge "J". The main jet drops away from the body of the carburettor as shown at "A."

be low and reducing further, but the engine will not be slowing down because the mixture is weak.

The same car going along a fast road at high speed in the same gear and with the throttle wide open, will have the piston somewhere near its maximum travel. This is because the engine suction will be high due to the increased rpm of the engine under less loading.

For cold starting the piston has the set minimum height between it and the top of the bridge so that air/fuel can be drawn into the engine for starting (otherwise there would only be a gap when suction was produced by cranking the engine over, which would lift the piston). The butterfly is also opened slightly when the choke mechanism is bought into operation and the fuel is temporarily enriched by the lowering of the main jet on most SUs. Once the engine is warmed up, the choke is pushed back in which restores the top of the main jet to its normal operating height, the extra amount that the butterfly is held open is also removed.

HIF-type carburettors don't use this choke system, instead they admit extra fuel to the engine using a separate fuel circuit. The result is the same.

Chapter 2
Inlet manifolds

The inlet manifold is often one of the least considered parts of the induction system. All too often an engine will have all of the 'right' bits on it but not perform as well as might be expected. A possible reason for this lack of performance can often be traced to an inlet manifold which is now restrictive. The carburettors may well have been changed to larger ones but, in many instances, the inlet manifold is virtually the same size internally as it was before. Enlarging the runner area immediately next to the carburettor mounting flange is often all that is done and, usually, this is not enough. The engine's power potential will not be realised. Remove the possibility of the inlet manifold being a performance limiting factor by checking that it has suitable internal dimensions.

Inlet manifolds must be thought of as extensions of the cylinder head inlet port. It's quite common to find a cylinder head with ports that have been opened up being fed air through an inlet manifold with runners that are much smaller in cross-sectional area. As a general rule, the way to check to see whether the size of the inlet manifold runners is reasonable for the particular engine is to measure the diameters of the cylinder head inlet ports and make the inlet manifold runners at least 0.040-0.080in/1-2mm larger in diameter (for other than round ports, measure the cross-

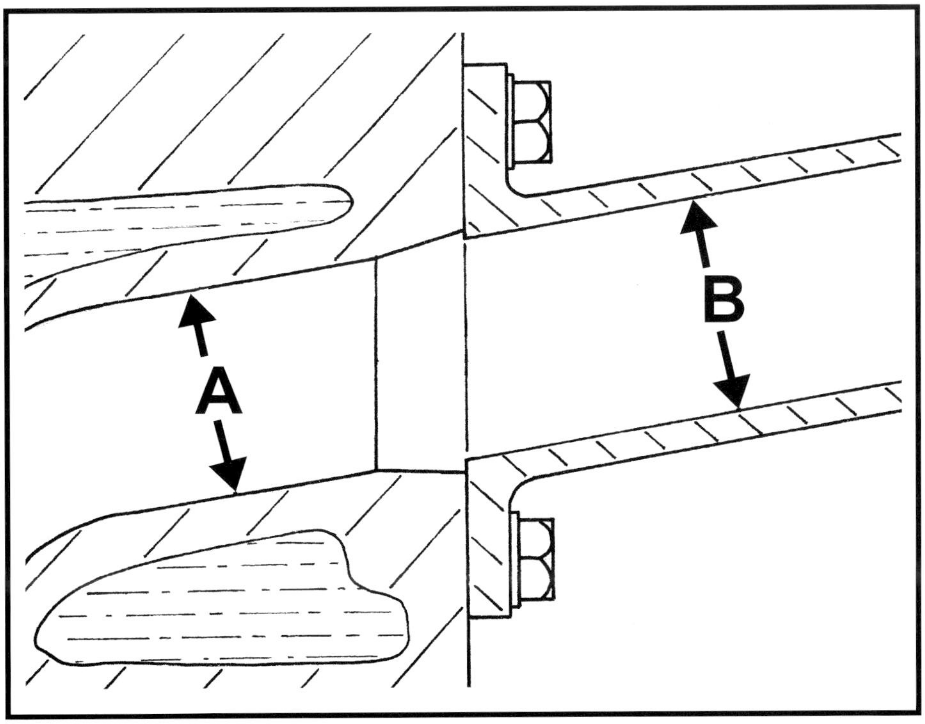

With round inlet ports, the round inlet manifold runner should be larger at point "B" than point "A" of the port.

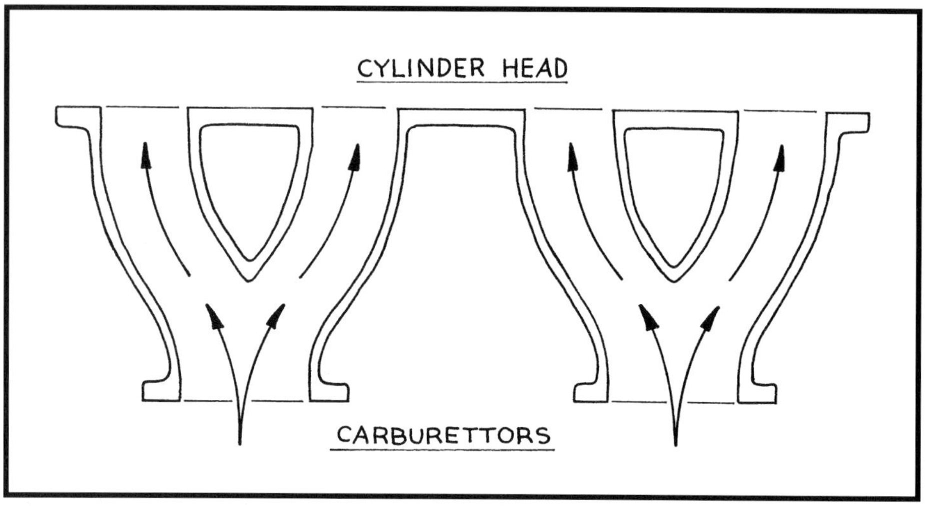

Cross-section of a typical twin SU inlet manifold to suit a four cylinder engine with four individual inlet ports in the cylinder head.

sectional area). The cylinder head ports should **not** be measured at the inlet manifold matching face because many cylinder head porting businesses open up this area considerably. The inlet ports should be measured 1-1.5in/25-38mm into the port to get a more realistic appraisal of true size (the ports will almost always be smaller a little way into the actual port). Whether the inlet port be round, square or rectangular the inlet manifold runners should, at the very least, be the same size in cross-sectional area (if not slightly more) as the measurements found in the cylinder head inlet ports.

The suitability of an inlet manifold for your individual application is actually quite easy to assess, simply by measuring the internal passages of the inlet manifold and then, if necessary, porting the manifold to enlarge the cross-sectional area of the inlet manifold runners.

There are two elements to inlet manifold runner sizing. One is making sure that where the inlet manifold diameter it meets the carburettor is at least the same as, if not slightly more than, the carburettor throttle bore diameter and that the two line up with each other (this is easily checked by bolting the carburettor to the inlet manifold and, with the butterfly fully open, looking through the throat of the carburettor with the aid of a torch: the inlet manifold must not overlap the carburettor's throttle bore). The second is making sure (and altering, if necessary) the inlet manifold runners are at least equal to the ports in cross-sectional area, as previously described.

Using SUs on a high performance engine almost always means one carburettor feeding two cylinders and, as a consequence, the inlet manifold runner splits into two individual runners some distance from the carburettor mounting flange. The split takes place very close to the carburettor mounting flange and it is not easy to measure this transitional part of the manifold because the internal shape and size usually starts altering within the first 0.25in/6mm into the inlet manifold.

There are several scenarios when it comes to measuring port and runner sizes because some engines have square inlet ports and manifold runners, some have round inlet ports and manifold runners and some have rectangular inlet ports and manifold runners. The round port runner engine's inlet manifold and cylinder head configuration is the easiest to deal with as the measuring and checking is done on a straightforward diameter basis.

With regard to round port/runner engines, the best way of checking to make sure that your particular inlet manifold has suitable internal dimensions is to measure the cylinder head inlet port diameter using inside callipers (then measuring across the calliper tips with a vernier calliper). This method will give a diameter size against which the inlet manifold can be checked.

Take it that a well modified engine fitted with one SU carburettor for every pair of cylinders will require individual inlet manifold runners with approximately the following diameters:

250cc per cyl: 1.100-1.140in/28-29mm
300cc per cyl: 1.140-1.180in/29-30mm
350cc per cyl: 1.180-1.220in/30-31mm
400cc per cyl: 1.220-1.260in/31-32mm
450cc per cyl: 1.260-1.300in/32-33mm
500cc per cyl: 1.300-1.340in/33-34mm
550cc per cyl: 1.380-1.420in/35-36mm
600cc per cyl: 1.420-1.455in/36-37mm
650cc per cyl: 1.455-1.495in/37-38mm
700cc per cyl: 1.495-1.540in/38-39mm
750cc per cyl: 1.540-1.575in/39-40mm

It's worth noting that the E-Type XK Jaguar engine of the 1960s was fitted with three 2in SUs and the individual inlet manifold runners were approximately 1.535in/39mm in diameter internally. The standard inlet manifold on this particular engine never needed the runners to be enlarged as the internal sizes were generous. Take it that this example represents the largest displacement per carburettor ever likely to be encountered. These XK engines are of 3.8 litre and 4.2 litre capacity (individual cylinder capacities of 635cc and 700cc respectively), with each SU feeding two cylinders. The 2in SU is the largest SU carburettor made and, make no mistake about it, when these XK engines are fully modified to an 8000rpm/300bhp capability, the 2in SUs and the standard inlet manifold are often still in place (modified a little perhaps, but still completely recognisable as SUs).

Not all inlet manifolds can be ported out to the suggested sizes, but most can. In cases where inlet manifolds can't be taken out to the ideal size it should still be ported out as much as possible without weakening the structure. Even if the inlet manifold's individual runner sizes end up less than ideal, if they are opened out as much as possible the engine will definitely go a lot better.

Take it that 0.080-0.100in/2-2.5mm manifold wall thickness is the minimum to have on the basis of good overall casting strength if the individual inlet manifold runners are unsupported. Measure the outside diameter of the inlet manifold with a vernier calliper and deduct the 0.197in/5mm from this size to deduce the maximum internal bore size. Many cast aluminium inlet manifolds do have supporting sections on them which means that the individual 'runners' are not structural or not entirely structural. Inlet manifolds like this can be ported out to a wall thickness of 0.080in/2.0mm and the casting will not normally break.

Any inlet manifold which is found to have individual runners that are too small and has insufficient material wall thickness to be taken out to the suggested sizes is going to have to have strengthening gussets added before it is ported out, or you may be able to find an alternative manifold. If wall thickness in any area is too thin after porting, that part of the inlet manifold will have to be welded to strengthen it or fill any holes if the manifold has been ground through. This may appear to be a lot of work but it is still easier and less expensive than fabricating an inlet manifold from scratch. Most inlet manifolds are relatively easy to rework (fortunately).

If the inlet manifold is restrictive power will be down. Of course, if the cylinder head has been over-ported, measuring of the cylinder head inlet port diameter will not be a very good guide to manifold runner size. The engine capacity/runner size table does, however, address this eventuality by giving a maximum individual runner size.

If a particular inlet manifold cannot be ported out to anything like the required size, an alternative inlet manifold must be found or, failing that, an inlet manifold will have to be fabricated. Essentially, making inlet manifolds is not within the remit of this book and, as a consequence, is not being recommended. What is being recommended is to maximise the potential of the inlet manifold that you have. This may mean welding on strengthening gussets and building up certain areas to maintain a uniform inside diameter.

The problem of finding enough wall thickness to open runners out to the optimum size is a reason why SUs and their manifolds are replaced by Webers/Dellortos. Factory fitted twin SU carburettor systems were almost always made for relatively standard engines with inlet manifolds to match, whereas Weber/Dellorto sidedraught inlet manifolds were usually made for well modified engines from the very start, and therefore have larger manifold passageways, or the casting is substantial enough to be ported out considerably.

The easiest way of checking a round manifold runner's individual diameter (after working out what size is needed) is to make an old engine valve into a gauge. This is actually the most simple of gauges to make and is usually ideal for checking the inside diameter of an inlet manifold runner. The diameter of the head of the valve is machined to the required diameter and the stem of the valve can then be used as a handle to insert the head of the valve into the runner. Valves come in many diameters and any valve head can be turned down - even exhaust valves can be turned down using a tungsten carbide turning tool. Inlet valves are easier to turn, of course, but are generally larger in diameter. Inlet valves are also normally turned down with a tungsten carbide tool, but a high speed steel tool will usually be quite sufficient.

With the required internal size of the finished inlet manifold runner known, two gauges will be required to measure the porting work. If, for example, the finished size of the inlet manifold runners is to be a diameter of 1.340in/34mm and the inlet manifold is originally 1.180in/30mm, have one gauge sized to 1.280in/32.5mm and the other to 1.320in/33.5mm. Use the smaller gauge for the initial porting work and carry on removing material until the gauge will pass completely through the runner; then use the second, slightly larger, gauge to check progress during the second stage of porting to finished size. The smaller gauge is used to get the bulk of the material out of the runner, while the larger gauge is used to carefully finish the runners to final size. The diameter of the finished runner (about 1.340in/

34mm) will actually be a little larger than the gauge because, to pass the larger gauge through the runner, some clearance is necessary.

While the examples described so far have related to round port/runner engines, the same basic principles apply to square or rectangular types. It's a bit more difficult to check the sizes of square/rectangular runner inlet manifolds, some actually being impossible to check. The standard Rover V8 inlet manifold, for example, can't be checked throughout or resized. Any square/rectangular port cylinder head and inlet manifold can usually be measured using inside callipers or, preferably, a gauge made out of 16 gauge sheet steel with a stem brazed onto it. Forget methods like 'eyeballing' which are nothing more than guesswork, with which it is just too easy to make a mistake. Direct measurement is the only method to use, with the gauge method being superior to using inside callipers.

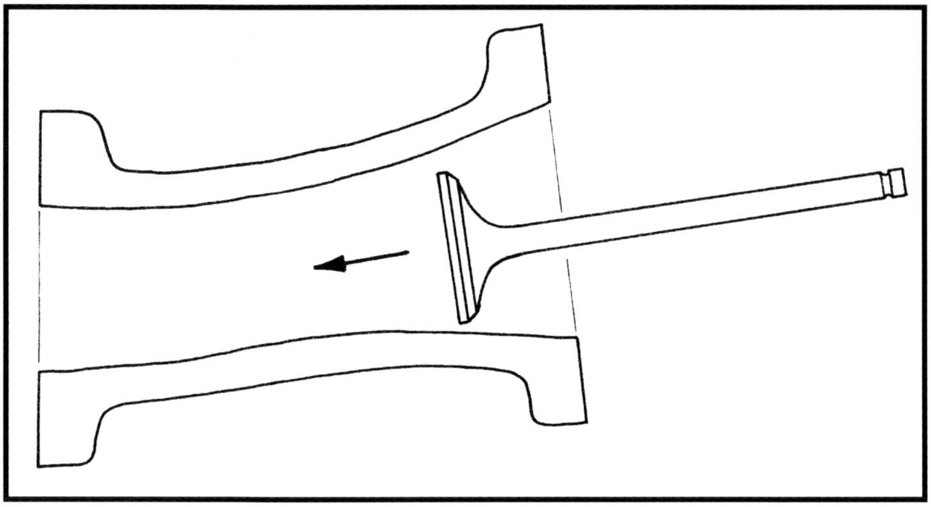

Turned down valve head being used as a gauge to size the inlet manifold runners. The valve's head must pass right through the runner.

Chapter 3
What size carburettor/s?

The sizing of the carburettors is, in the overall scheme of things, matched to the airflow demand of a particular engine. The demands of a modified engine are often far in excess of the standard engine. An engine's power output is related to how much air can be moved through it, and airflow capacity of the carburettion must be matched to the engine's optimum airflow capability. The only way to improve the power output of an engine that has insufficient carburettion is to fit a larger carburettor, or carburettors.

If the air/fuel mixture is correct and the carburettors are in fact limiting engine performance because they are at maximum airflow, the engine will accelerate well but will abruptly reach a rev limit and hold the rpm. This will apply to top gear performance as well, unless the vehicle is over-geared. In this scenario it is assumed that the inlet manifold is not the restrictive factor and that it has been sized to suit the head's inlet ports.

If an engine has two 1 1/2in SUs and power is limited by the fact that the carburettors are too small, the solution is to fit two 1 3/4in SU carburettors. As long as the engine can actually use the increased airflow capacity provided by the larger carburettors, the engine power will increase. If however larger carburettors are fitted but the inlet manifold runners are not enlarged to suit the size of the carburettors' throttle bore diameter/cylinder head port sizes, there may be no gain in engine power except perhaps at very high rpm. On road going vehicles it is often better to fit carburettors on the smaller side of the carburettor size range, whilst going for larger carburettor sizing if the vehicle is mainly for track use (constant high rpm).

If an engine has a single 1 3/4in SU and twin 1 1/2in carburettors on an appropriate inlet manifold are substituted, there may well be mixed results, especially if the airflow capacity of the single carburettor is sufficient for the engine to attain maximum rpm. The air/fuel distribution may well be better with the twin carburettors but, because the airflow capacity of the single carburettor is well matched to the airflow capability of the engine, it only needs the inlet manifold to have a large internal cross-sectional area and the engine will almost always perform nearly as well, or equally as well, as the same engine fitted with the twin carburettor set-up.

SU CARBURETTOR SIZES

SU have made a considerable number of carburettor types over the years, but the butterfly sizes of the carburettors were limited to six. The sizes range through 1 1/8in, 1 1/4in, 1 1/2in, 1 3/4in, 1 7/8in and, finally, 2in. The range of butterfly sizes has encompassed the entire type range with no one carburettor type using all six butterfly sizes. As it happens, this doesn't matter, so choose the carburettor size from the type range and use the number of carburettors that suits your engine. It is however customary to have one SU feeding two cylinders on high performance engines.

SU actually offer a good range of sizes and the only size that does not need to be considered is the H1 or 1 1/8in butterfly model as this is now a very old carburettor and the 1 1/4in SU is universally used in its place. However, if this is the size of carburettor you have and you want to use the tuning methods described in this book, the procedures for the H1 are the same as for any other SU.

SU TYPE RANGE

H1: 1 1/8in diameter butterfly
H2: 1 1/4in diameter butterfly
H4: 1 1/2in diameter butterfly
H6: 1 3/4in diameter butterfly

HD4: 1 1/2in diameter butterfly
HD6: 1 3/4in diameter butterfly
HD8: 2in diameter butterfly

HS2: 1 1/4in diameter butterfly
HS4: 1 1/2in diameter butterfly
HS6: 1 3/4in diameter butterfly
HS8: 2in diameter butterfly

HIF4: 1 1/2in diameter butterfly
HIF6: 1 3/4in diameter butterfly
HIF7: 1 7/8in diameter butterfly
HIF38: 38mm/1.496in dia. butterfly
HIF44: 44mm/1.730in dia. butterfly

For high performance applications avoid using an SU which is fitted with an 'overrun valve' in the butterfly. Either buy a new butterfly disc without the valve, or remove the basic mechanism from the original butterfly disc and solder-up the resulting hole. Use only manual choke model carburettors for high performance applications.

Take it that a 1 1/4in SU can feed a standard 1000cc four cylinder engine, a 1 1/2in SU a standard 1800cc four cylinder engine and a 1 3/4in SU a standard 2400cc four cylinder engine. This assessment is reasonable for the cylinder capacity of

An "overrun valve" in a butterfly is a considerable obstruction to airflow.

these engines in view of the fact that most standard engines are very often required to rev to 4000-5000rpm only and that the torque characteristic of the engine, coupled with fuel economy, is more important than the maximum power potential for standard road use.

Considering that the engines in question are modified for high performance, the sizing situation can be broken down to the following:

SINGLE CARBURETTOR APPLICATIONS (FOUR CYLINDER ENGINES)

A modified 850-1000cc four cylinder in-line engine fitted with a single carburettor can use either a 1 1/4in SU or a 1 1/2in SU.

A modified 1000-1300cc four cylinder in-line engine fitted with a single carburettor can use a 1 1/2 or 1 3/4in SU.

A modified 1200-1500cc four cylinder engine fitted with a single carburettor is going to need, at the very least, a 1 1/2in SU and more likely a 1 3/4in SU and a heavily modified 1500cc engine will very likely use a 1 7/8in SU.

A modified 1500-1700cc four cylinder in-line engine fitted with a single carburettor can use a 1 3/4 or 1 7/8in SU.

A modified 1700-2000cc (or more) four cylinder in-line engine fitted with a single carburettor can use a 1 3/4, 1 7/8 or 2in SU.

TWIN CARBURETTOR APPLICATIONS (FOUR CYLINDER ENGINES)

Twin SUs are commonly found on high performance four cylinder in-line engines, but there have been plenty of six cylinder in-line engines that had twin SUs as standard equipment. Twin SUs could be used on any four cylinder engine configuration.

A modified 850-1200cc four cylinder engine fitted with twin

carburettors is going to need two 1 1/4in SU carburettors. A 1000-1200cc engine will need to be heavily modified to justify 1 1/2in SUs, with most using 1 1/4in SUs. A heavily modified 1200-1400cc four cylinder engine is going to need two 1 1/2in SUs.

A modified 1200-1600cc four cylinder engine is going to need two 1 1/2in SU carburettors, while a heavily modified 1500-1600cc unit might well be able to use two 1 3/4in SUs.

A modified 1600-1800cc four cylinder engine is going to use either two 1 1/2in SUs or two 1 3/4in SUs. Most 1600-1800cc engines will use 1 3/4in SUs. For a high rpm application and an engine having a well ported cylinder head the 1 3/4in SUs will prove the better size, the 1 1/2in SUs going as well as the larger carburettors up to the limit of the carburettor efficiency (wherever that point may be).

A modified 1800-2300cc four cylinder engine is going to use either two 1 3/4in SUs or two 1 7/8in SUs. Because of the easy availability of 1 3/4in SUs compared to 1 7/8in SUs, the 1 3/4in SUs are likely to be tried in the first instance. Any engine that is equipped with 1 3/4in SUs as standard and then has a pair of 1 7/8in SUs substituted will almost certainly need to have the inlet manifold well ported out to make the best use of the larger carburettors. If the individual inlet manifold runners are not altered there may well be only a very small improvement in engine performance, if any.

A well modified 2300-2500cc four cylinder engine is going to need 2in SUs.

There is some overlap with carburettor sizing on four cylinder in-line engines. This is because of the relatively small individual cylinder capacities. The way to get the sizing right first time is to be very realistic about how heavily modified the particular engine is, and whether it is primarily for road or racing purposes. For road use, fitting smaller carburettors from the appropriate size range seldom results in poor overall engine performance.

TRIPLE CARBURETTOR APPLICATIONS (SIX CYLINDER ENGINES)

Triple SUs are commonly found on high performance six cylinder in-line engines. Most of the more common six cylinder engines have capacities of between 2000-4500cc. The carburettor size range for use with these engines is 1 1/2in, 1 3/4in, 1 7/8in and 2in.

A 2000-2800cc six cylinder engine is going to need three 1 1/2in SUs, while heavily modified units will need 1 3/4in SUs.

A 2800-3800cc engine is going to need three 1 3/4in SUs, while heavily modified units will be likely to need 1 7/8in SUs.

A 3800-4500cc six cylinder engine is going to use three 2in SUs.

QUADRUPLE CARBURETTOR APPLICATIONS

The sort of engine that could use four SUs is either a V8 or a V12. An engine for which a quadruple SU set-up is available as a conversion is the British built Rover V8. This carburettion system uses four original equipment 1 3/4in SUs on a steel fabricated inlet manifold and was developed along these lines to utilise the original two carburettors. It works very well and offers much better fuel distribution than the original equipment inlet manifold. The inlet manifold runners are however only just the right size. The inlet manifold main runners are made from rectangular section, thin walled, mild steel tubing which is commercially available.

Approximate carburettor sizes related to individual cylinder capacity and one carburettor feeding two cylinders:

1 1/4in SU: 200-250cc per cylinder.
1 1/2in SU: 250-400cc per cylinder.
1 3/4in SU:400-500cc per cylinder.
1 7/8in SU: 500-600cc per cylinder.
2in SU: 600-700cc per cylinder.

CARBURETTOR CHOICE COMPROMISES

The situation is always going to arise where a carburettor, or carburettors, are available which are slightly larger than ideal for the size of engine. The inevitable question is going to be should they be used? A **small** mis-match is no cause for concern, but the fitting of carburettors which are near the recommended size change point is often another matter. The situation is always going to revolve around how heavily modified the engine in question is, because this relates directly to the airflow requirement.

The suggested carburettor sizes given in this chapter will be correct in most cases, but a heavily modified engine is definitely going to need the larger carburettor sizes in the recommended range, while more lightly modified engines will work best with the smallest carburettor in the range.

Chapter 4
Needles

There are three basic needle diameters used in SU carburettors. Needles are termed "90s," "100s" or "125s." The 90 suits a 0.090in diameter main jet, a 100 suits a 0.100in diameter main jet and a 125 suits a 0.125in diameter main jet. The 1 1/4in SUs, 1 1/2in SUs and later HIF38 SUs all have 0.090in diameter main jet sizes. The 1 3/4in SUs, HIF44 SUs and the 1 7/8in SUs all use 0.100in main jet sizes. The 2in SU has a 0.125in diameter main jet size.

There are two types of SU needle. One needle type is for the earlier designs (H and HD) and is termed "fixed type" because it is firmly fixed into the piston. The later needle type is for the emission control type of carburettor (HS and HIF) and is spring-loaded with a completely different clamping method in the piston.

All needles are fitted in the piston of an SU carburettor in exactly the same relative position (the shoulder of the needle flush with the bottom of the piston). However, needles are model specific. The 0.090in and 0.100in needles are either "fixed" or "spring-loaded" types, while the 0.125in needles are always of the "fixed" type.

The criteria for selecting needles revolves around the fact that you do not have to measure the diameter of an actual needle but, instead, simply use the code stamped on it to identify it. The code is then looked up in the SU needle book which gives diameter measurements at the standard measuring points for *all* needles. A richer or leaner needle is found by cross-referencing the listed sizings.

The official SU needle diameter measuring points start at the shoulder and each subsequent measuring point is at 0.125in/3.2mm increments downwards towards the point of the needle. In order to choose one needle over another, it's essential to know what each measuring point diameter is. This is because each measuring point is 'zoned' in relation to what that particular area of the needle does in each size of carburettor.

The fact that the official *SU Carburettor Needle Profile Charts* book lists all needles and gives their diameters at the various measuring points, provides an exceedingly accurate way of working out the characteristics of one needle in relation to another. Needle choices are made on the basis of going richer or leaner (smaller or larger, respectively, at a particular standard SU measuring point) at any point in the rpm range. This may still mean trying out several needles before the 'right' one is found. This trial and error part of the procedure is what often causes confusion. The way to eliminate the confusion is to work out what diameters need to be changed after each needle has been tried in order to build a 'picture' of the ideal needle. Essentially you build on the results given by each needle tested.

The problem here is that needles are not particularly cheap and there may also be a small delay in getting alternative needles. The consolation is that once the needle specification is optimised the engine tune will be 'dead right.'

Money can be spent on a needle, or needles, which in five minutes

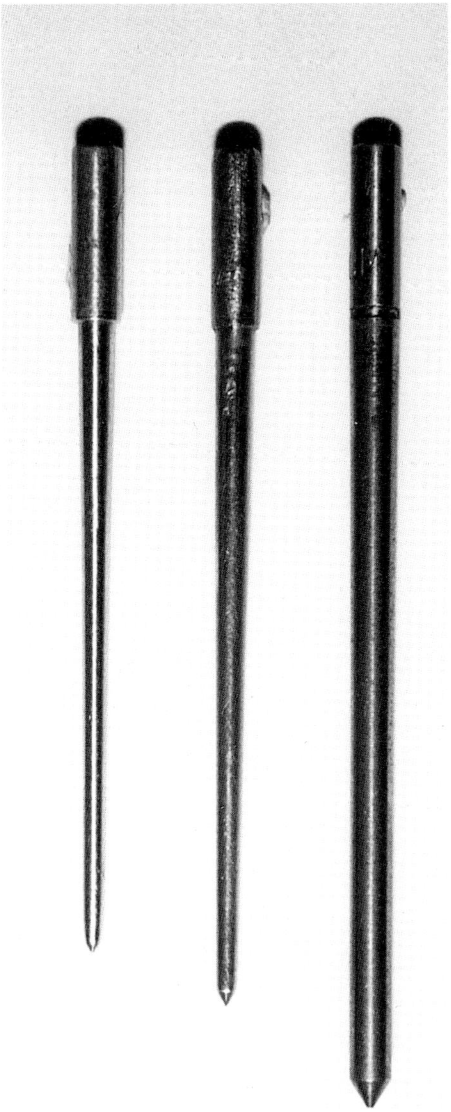

0.090in diameter needle on the left, 0.100in in the middle and 0.125in fixed needle on the right.

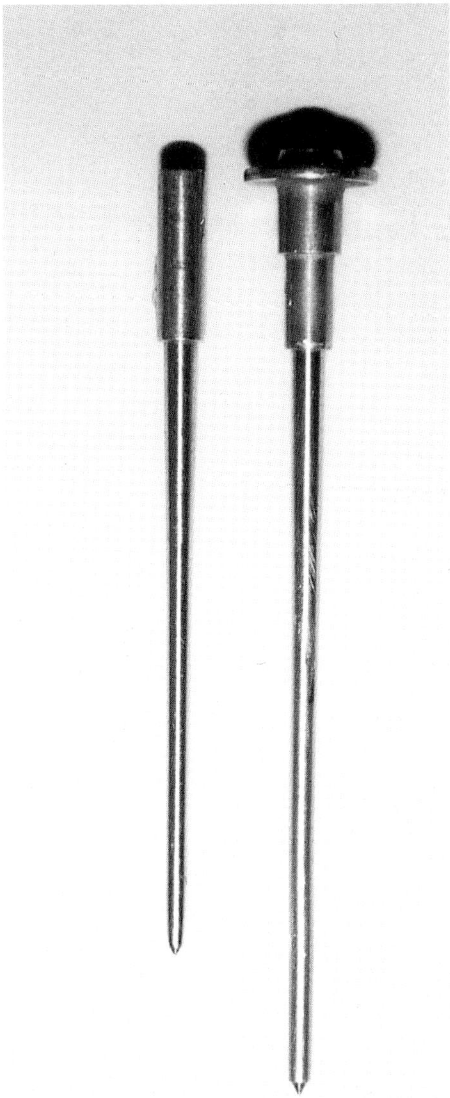

Fixed type of needle on the left and spring-loaded type on the right.

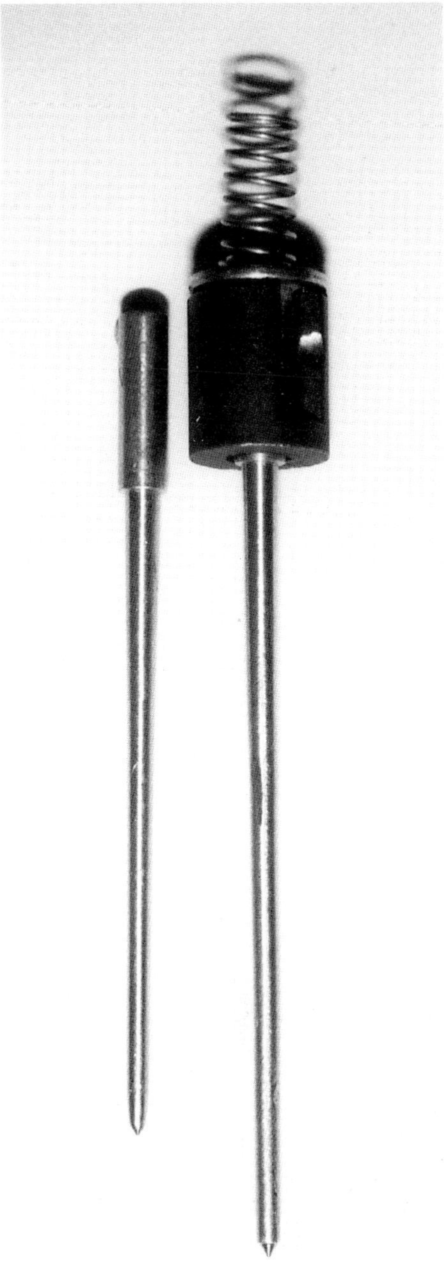

Fixed-type needle on the left fits into the piston just as it comes. Spring-loaded needle needs a spring and collar, as shown on the right, before it can be fitted into the piston.

prove to be incorrect for the particular application, more needles then have to be procured and this process can be frustrating. Unfortunately, there is no alternative to a little of this experimentation, but time and money can be saved by making sure that each subsequent size of needle bought is definitely a step in the right direction: in this way the choice of needles is quickly narrowed down. There is no point in haphazardly trying different needles in the hope that a particular size might work: with SUs, logic **must** be applied to the way in which needles are chosen.

Especially if more appropriate needles were not immediately available, the trend in the past (assuming greater richness was required) was to polish down the

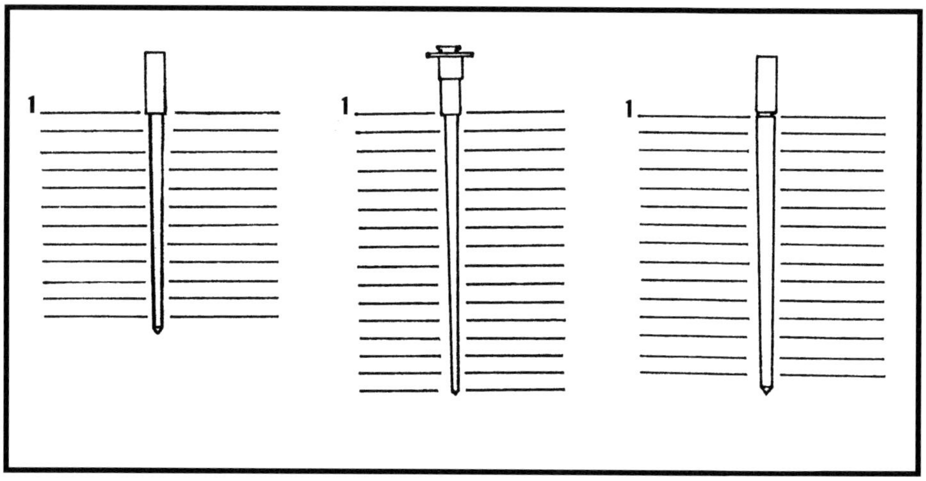

First measuring point position for *all* SU needles is at "1."

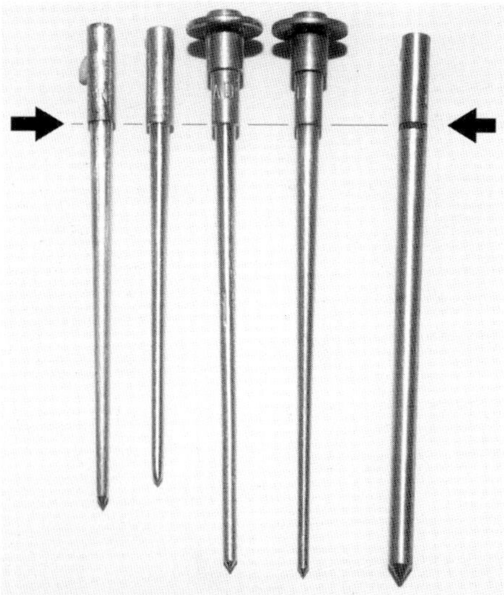

The first measuring points (arrowed) of five different SU needles.

existing needles for their full lengths with, perhaps, more emphasis on polishing towards the tips. Unfortunately, this process was almost always haphazard and hardly ever gave the desired results - unless you were very lucky!

In fact, there is nothing wrong with polishing SU needles to a new diameter - provided the end result is dead accurate. It is possible to polish a needle to a required size and make it 'richer' at any particular measuring point. This process can save time and money, but is only going to work if the needles' diameters are accurate: this is where the problems come in (getting one needle right and then making one, two or three identical copies!). There is no doubt that factory machined needles are ultimately the easiest option, but polishing needles can be of use to discover how far diameters can be reduced before the needles become unsuitable because the air/fuel mixture is becoming too rich. If a needle reduced in diameter by polishing has the right diameters at the right points for the particular engine application, the engine will go as well as that same engine equipped with a factory made needle of the same specification (if available).

You need to be aware that, if their history is unknown, your carburettors could be equipped with non-standard needles which have been polished down by a previous owner. This is why using brand new needles of known origin is the best starting point.

SU can't be expected to make and supply needles suitable for unusual applications (applications that they don't know exist). SU do make a comprehensive range of needles, but expect to have to 'make' your own needles for modified engines (see appropriate chapter for more information).

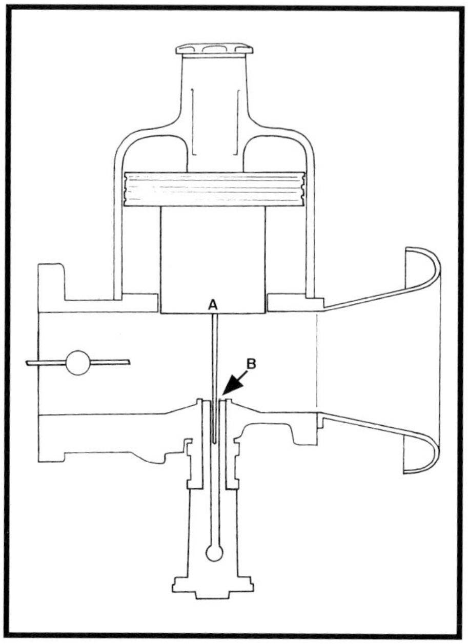

Piston "A" is at full lift, yet the point of the needle is well below the bridge. The 'active' area of the needle is from the underside of the piston to the top of the main jet "B" (which is always below the bridge).

Some SU needles are longer than others and all are longer than their working ranges (provided the needle is the correct one for the carburettor). The actual length - from shoulder to point - of 0.090in needles ranges between approximately 1.550in/ 39.3mm and 1.850in/46.9mm, 0.100in needles are approximately 1.915in/48.5mm long and 0.125in needles are approximately1.945in/ 49.4mm long. The overall length of a needle is not particularly relevant but, what is relevant is that area of the needle's length which represents the working range in the particular carburettor.

A piston in fully raised position (needle omitted).

Needle types are carburettor specific.

Each SU carburettor size has a needle of a certain length. However, only a portion of the needle length is actually used in metering the fuel supply between idle and full throttle. The area of the needle that meters fuel supply is from the shoulder downwards to a point before the tip of the needle. The active area can be determined by measuring the distance from the bridge of the carburettor to the base of the piston when the piston is at full lift: the measurement is made with inside callipers. Make this measurement on each carburettor to check that it is to specification and equal for all the carburettors on the engine. Note that some carburettors were restricted when manufactured by having the piston travel distance reduced to effectively make the carburettor smaller (when carburettors technically too large for the application were factory fitted to high performance engines). The piston to bridge sizes given in the following text are for unrestricted carburettors. If a restricted carburettor is paired up with an unrestricted carburettor the engine will not run correctly at the top end. Check all carburettors to establish that the carburettor pistons travel the same specified distance.

Piston to bridge dimensions (at full piston lift) -
1 1/4in = 0.0910in/23.1mm.
1 1/2in & HIF38mm = 1.125in/28.5mm.
1 3/4in & HIF44mm = 1.300in/33.0mm.
Later 1 7/8in = 1.520in/38.6mm.
2in = 1.715in/43.5mm.

The 1 3/4 HIF SU carburettor, for example, has a needle that is overall 1.900in/48.2mm long when measured from the 1st measuring point at the shoulder to the point. The 'working range' of the needle for this model of SU is, however, only approximately 1.340in/34.0mm (depending on how much the main jet has been wound down to effect idle). As it is the circumference of the needle at the top of the jet which determines fuel metering, the part of the needle below the top of the jet (approximately 0.520-0.560in/13.2-14.2mm) at full piston lift serves primarily to locate the needle in the jet.

The reason for the variation in the working range of the needle is the fact that the top of the main jet can be anything up to 0.080in/2mm below the bridge of the carburettor. The top of the main jet is the effective point from which the fuel is metered, not the bridge of the carburettor. The main jet is invariably wound down to a certain extent. The distance between the top of the main jet and the bridge is going to depend on the fuel requirements of the engine, commensurate with the needle size, at idle. To account for this variable factor all needle sizings to the measuring point beyond the measured height from bridge to piston (piston at

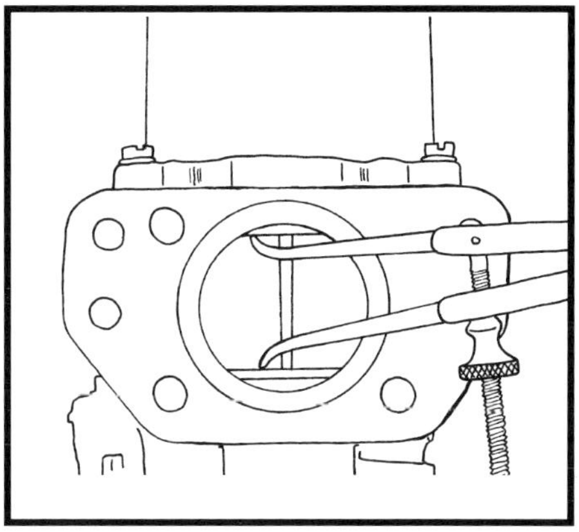

Distance from the bridge of the carburettor to the underside of the piston (with piston at full lift) is measured with a pair of inside callipers. Use a vernier calliper to get the measurement from the inside callipers.

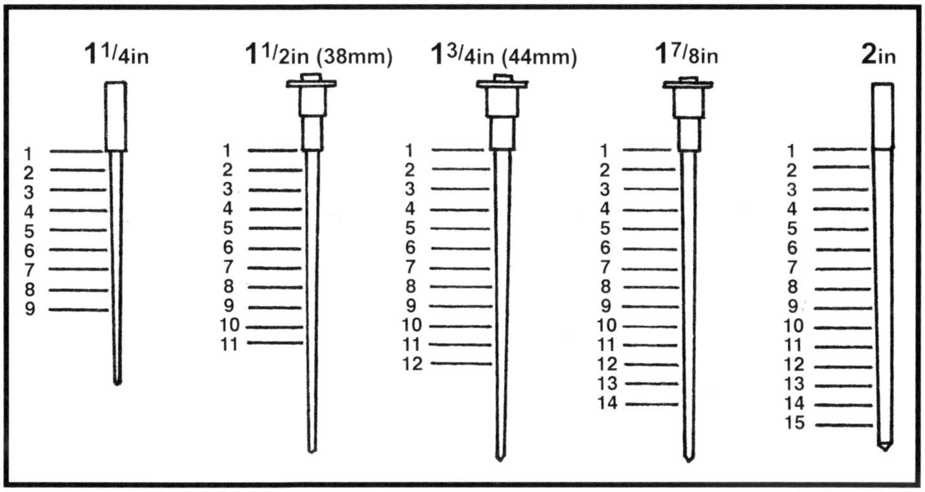

The needle's working range (from the shoulder) is equivalent to the distance from the bridge of the carburettor to the underside of the piston (when the piston is at full lift), plus whatever distance the top of the main jet is below the bridge.

This is the standard fitted needle position for *all* SU carburettors.

full lift) must be taken into account. The working range of the needle should always be assumed to be 0.080in/2mm more than the bridge to piston dimension to allow for main jet setting.

Although SU provides measurements for needles all the way down to the point, the tip area of the needle's length is a transitional section which does not have any real bearing on the effective fuel metering diameter. Using the 1 3/4in SU as an example to relate the working range of the needle to the SU factory needle drawing, the distance from the 1st point to the 12th point is the approximate working range. From the 12th point to 16th point (end of the needle) is the transitional part. For sizing purposes (alterations in the diameter of the needle at various points for fuel richness/leanness) the area between the 1st point and the 12th point is what is important. By going to the 12th measuring point variable main jet height settings are accounted for.

The fitted needle position for all SUs is basically the same: the shoulder of the needle should be flush with the bottom edge of the piston surface.

A very slight overall mixture strength increase is achievable with fixed needle type SUs by positioning the needle so that the shoulder is recessed into the piston. This can, in certain instances, make a slight difference to engine performance.

The working range for the needle in each type of SU carburettor including allowance for main jet adjustment -:

1 1/4in SU - up to the 9th measuring point

1 1/2in SU - up to the 11th measuring point

1 3/4in SU - up to the 12th measuring point

1 7/8in SU - up to the 14th measuring point

2in SU - up to the 15th* measuring point

* Note that with the 2in SU the 16th measuring point may have to be used if the main jet is wound down 0.080in/2mm from the bridge. The reason for this is that the 15th measuring point position may be 0.040in/1.0mm, or more, above the top of the main jet in this circumstance.

In the first instance the method of choosing a needle for a high performance application revolves around selecting a needle which is 0.001in smaller than the main jet diameter at the first measuring point. This reduces the total number of needles available for consideration. This applies to *any* needle being used in *any* SU. Some needles are made the same size at the first measuring point as the main jet diameter (or 0.0005in smaller than the main jet diameter): these needles always give an overall lean mixture and while they are excellent for their intended purpose, are seldom suitable to a high performance application.

It is possible to find several needles with the same 1st measuring point and full piston lift measuring point diameters, but with quite different mid-range measuring point diameters. No two needle specifications are identical, but many needles are very, very similar.

SUMMARY

The significant point of this chapter is that each size of SU carburettor has a certain amount of vertical piston movement and, while SU needles come in various lengths, only a certain segment of the needle is ever used to meter fuel.

Chapter 5
Altering needles

Although a vast range of SU needles is available, often the standard needle which best matches your ideal specification still has a couple of measuring point diameters which are not ideal. In the worst cases, even if nearly all measuring point diameters are wrong, it is quite possible to alter the needle provided it is larger than required at all appropriate measuring points. Tailor-making a needle requires a method which allows polishing of the needle in a specific position and a rate of material removal which is not too quick. All SU needles are factory-sized using a set method of dividing needle length into increments of 1/8 inch from the shoulder (1st position) down to the point of the needle.

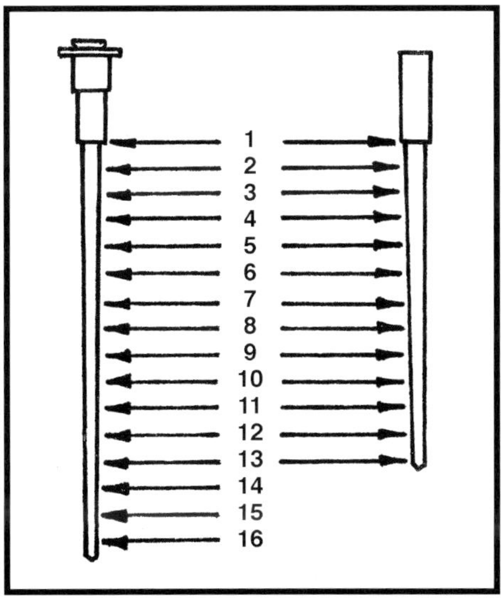

Factory designated measuring points for *any* SU needle.

METHODS OF HOLDING NEEDLES DURING POLISHING

The way to remove material from a needle is to polish it while it is spinning at around 200 to 400rpm. This process is best accomplished by mounting the needle in a drill press chuck.

There are two types of SU needle: "fixed" and "spring-loaded." The fixed type of needle is easy to hold in a chuck and presents no problems at all. On the other hand, the spring-loaded needle does present a problem in how to hold it easily in a machine tool chuck. The most accurate way of holding any spring-loaded SU needle is to make a split collar from a piece of mild steel round bar. One collar will suit both A type/ 0.090in and B type/0.100in spring-

'Fixed' type of needle being held in a drill chuck.

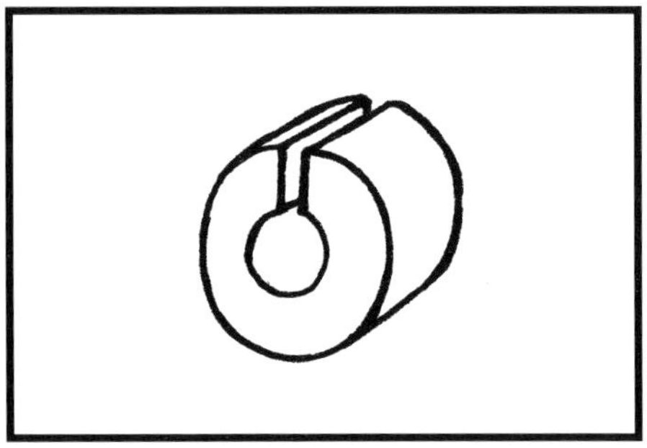

Split collar made from mild steel round bar.

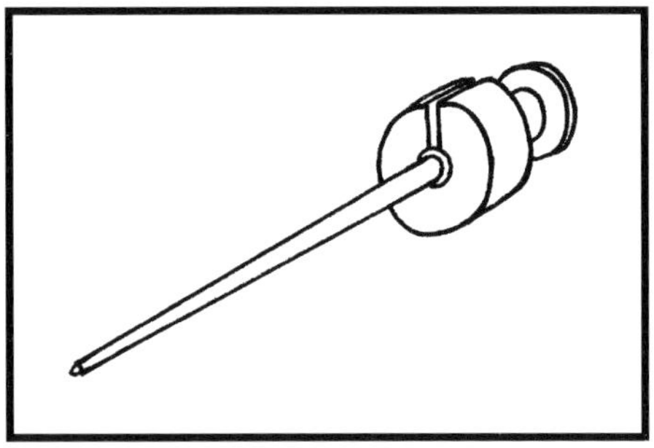

SU needle mounted in split collar.

loaded needles. The dimensions of the collar are 0.345in outside diameter, by 0.240in long, with a drilled and reamed inside diameter of 0.125in. Use a junior hacksaw to cut a lengthwise slot completely through one wall of the collar. With the needle fitted in the split collar, the collar itself can be held in the chuck of a drill press. The clamping pressure of the chuck will be enough to cause the collar to tighten around the needle and clamp it firmly. The needle is held accurately if the collar has been accurately made.

Another way of holding a spring-loaded-type SU needle is to use rubber tubing to act as a guide when the needle is fitted in the drill press chuck. The rubber tube needs to be 8.5mm in external diameter with a 4mm internal diameter and be 10mm long. While the chuck jaws will be clamping on the thin edge of the needle flange, the rubber tubing will

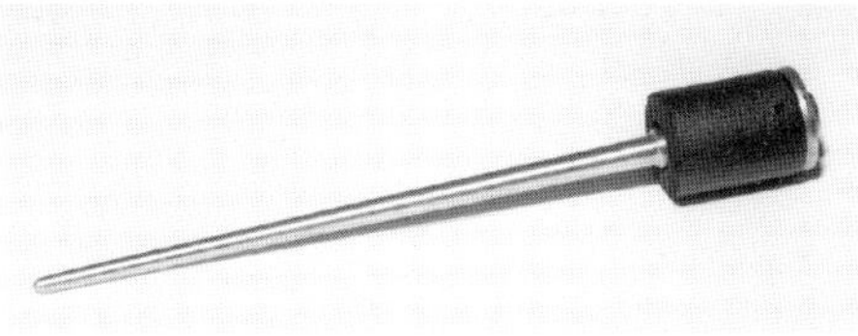

'Spring-loaded' type needle with a short piece of rubber tubing (over the two larger diameters above the measuring points) which assists with lining up the needle when it is held in a drill chuck.

'Spring-loaded' type needle being held in a drill chuck with the aid of a short length of rubber tubing to assist its alignment.

act as a good guide for reasonable alignment.

SU needles are almost always made from brass and are comparatively soft. It does not take too much effort to effect material removal.

MODIFYING INDIVIDUAL MEASURING POINT DIAMETERS

Warning! - The following procedure will bring your fingers into close proximity to a rotating chuck. Make sure you are not wearing loose clothing and proceed with due caution. If you are not completely confident that you can do this work safely, have an experienced engineer undertake it on your behalf.

180-220 grit wet and dry paper is used to remove the material from the needle and 600 grit wet and dry paper to finish polishing it. The method of polishing in exactly the right place revolves around accurately marking the non-abrasive side of the paper with lines 1/8in apart. This is done using a ruler and a ballpoint pen, then each line is numbered. The straight edge of the paper represents the 1st measuring point, the first line inwards the second measuring point and each successive line numbered accordingly. You should mark sufficient lines to cover the needle's working length.

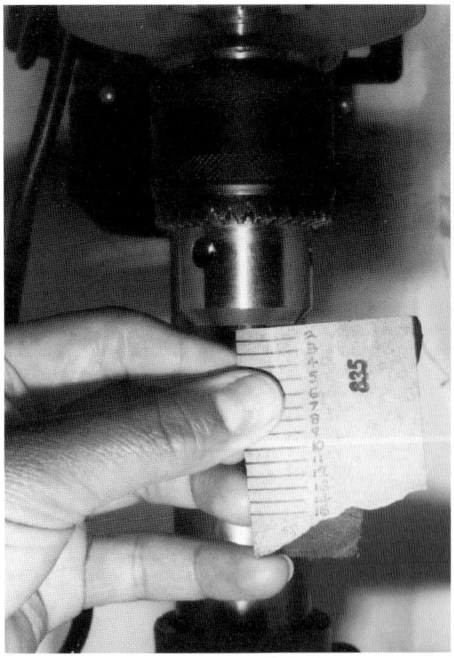

A short squeeze at each measuring point is all that's required to reduce the needle's diameter. Take care to ensure the abrasive paper does not move up and down in relation to the needle.

The wet and dry paper is then folded and wrapped around the needle. With the edge of the paper up against the shoulder of the needle, each measuring position is clearly identified. To remove material from any one position, finger pressure (thumb and index finger) is applied to the appropriate measuring point position. With a needle being rotated at 200 to 400rpm, five seconds' worth of finger pressure will remove approximately 0.001in from the diameter of the needle if the wet and dry paper is fresh and not clogged with material.

To check how much material is removed over a given period of time with your own set-up, select a measuring point and, working on a spare needle, apply finger pressure for exactly five seconds and then measure how much material was removed. You'll quickly develop a 'feel' for how much material you'll actually remove in five seconds.

The best way to accurately measure an SU needle at the various measuring points is with a micrometer. A vernier calliper is not as accurate as a micrometer, though some vernier callipers with digital readouts are very accurate and easy to use. Personal preference comes into the choice of measuring instrument and, as we are only looking for 0.0005-0.001in accuracy, use whichever instrument suits you best.

To serve as a gauge to position the micrometer or vernier calliper at a particular measuring point, use a piece of thin card (2 x 3in) with numbered lines 1/8in apart drawn on it (in the same way as the abrasive paper was marked). With the piece of card held under the needle and against its shoulder, all measuring points will be clearly shown.

While the methods described for

Card with the measuring points accurately drawn on it and then folded on the line.

The needle can then be positioned on the card with its shoulder (its first measuring point) against the edge of the paper.

removing material from the needle and identifying individual measurement points may appear to be a bit 'rough and ready,' they are in fact accurate enough to get any needle measuring point to within 0.001in or less of the size required. With this process practice makes perfect, but you'll find that the 'right' needle for the application can be made using these methods. Also, if more than one needle is required accurate duplication (needles being within 0.001in of each other) is quite possible.

It does take time and care to rework SU needles in this way, but there is often no alternative if you don't want to compromise engine performance. A final light polish of the

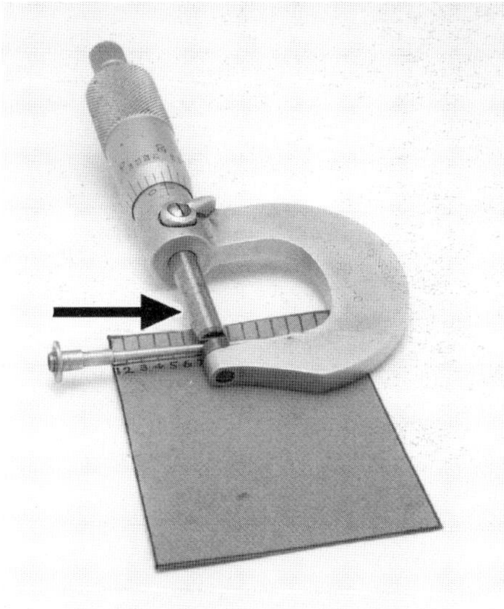

When measuring a needle measuring point, the front of the micrometer spindle (arrowed) is the datum.

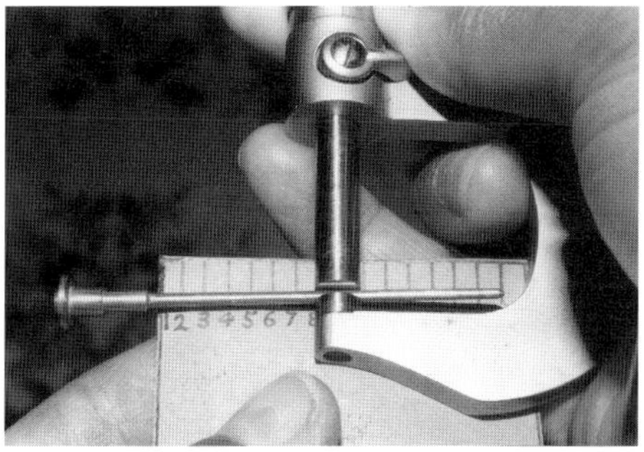

Needle being measured with a micrometer at the 8th measuring point. A piece of card with lines 1/8in apart indicates where to measure the needle. In practice, the micrometer is held in one hand and the card in the other. The needle is lightly gripped by the anvils of the micrometer and positioned against the card. Once this technique has been mastered, measuring is a 5 second job.

needle with 600 grit wet and dry paper effectively removes very little material but shines the needle surface to much the same finish as original.

The fact that the needles are perfectly round when made by SU means that when material is removed by the method described the roundness of a needle is basically

Card marked with measuring points is held next to the needle. Wet and dry paper is wrapped around the needle and, where material is to be removed, pressure applied to the abrasive paper adjacent to the relevant lines on the card.

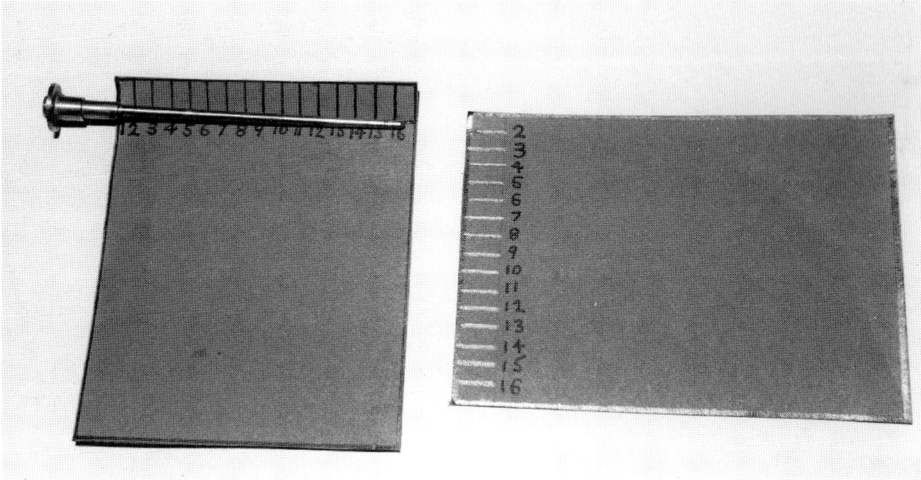

Two cards need to be made if the faster method of material removal is to be used. Card on the left is for measuring the needle diameters, card on the right is used for alignment. This layout suits right handed people.

maintained: ovality is possible, but not likely because of the usually small amounts of material being removed. Admittedly, the needle measuring point diameters nearer the point of the needle may end up a little oval but, effectively, they don't matter because they don't directly meter fuel.

Important: all measuring points after (towards the needle tip) the full throttle measuring point need to be progressively (0.001-0.002in on average) smaller down to the point.

Alternative method

Once you've mastered the method previously described, you can adopt a faster method of polishing needles. The second method involves holding a piece of card - marked with all the measuring points - adjacent to the needle (which is mounted as previously described) so that the first measuring point on the card is aligned with the shoulder of the needle. With the needle spinning, a small piece of wet and dry paper is wrapped around it and pressure applied at each relevant measuring point in turn.

Constant monitoring of the amount of material being removed is essential and a second card will be required for this purpose.

The alternative method of needle polishing is faster because it's very easy to move the abrasive paper around the needle after each squeeze, so that a fresh, unclogged, abrasive area is presented to the needle's surface. The first method, though more accurate for the beginner, requires constant refolding of the abrasive sheet as it becomes clogged.

SUMMARY

While it takes a certain amount of manual dexterity to reprofile an SU needle, practice does make near perfect. Patience will pay off: if you don't rush to remove material from the needle, the right measuring point diameters can be arrived at relatively quickly and without ruining the needle.

Chapter 6
Choosing starting point needles

Note: for racing applications read chapter 8 (which gives details of modifications that can be made to standard SU needles) before doing anything about ordering needles as described in this chapter.

***

SU carburettors will most often be used on engines that were originally fitted with SUs, or which are related to an engine that did, simply because appropriate inlet manifolds are readily available. As a consequence, the needles likely to be used as starting points for a given engine are easily identified because the *SU Reference Catalogue* lists the needle specification of every engine originally equipped with SUs.

Of course SU carburettors will work with any conventional four-stroke engine if appropriate manifolding is available or can be fabricated. If the engine unit you intend to use has never ever been fitted with SU carburettors as standard, go through the *SU Reference Catalogue* and look up the similar-sized engines with the same number of cylinders and the same number of carburettors. If the engine you intend to use is, say, an 1800cc four cylinder unit, go through the *SU Reference Catalogue* and list all of the engines around that size (1700cc to 1900cc) which have used the same number of carburettors as your application. With the needle and spring specifications from the various carburettors noted, make a chart of the needle sizes (from the 1st measuring point to the end of the working range) using the *SU Carburettor Needle Profile Chart* book. Choose the needles from this range which are the smallest in diameter (richest) from the 2nd to the 5th or 6th measuring point (needles that give low end and mid-range richness). This is not to say that needles chosen like this will be ideal for your application, but they will be a very good starting point. In fact, it is exceedingly likely that needles chosen in this manner will require only minor re-sizing at certain measuring points to obtain correct mixture strength. The reason for choosing needles which have the smallest diameters between the 2nd and 5th/6th measuring points is that this will usually mean that the engine's low end and mid-range air/fuel mixture will be on the slightly rich side (a likely requirement for a high performance unit). Also, if the engine's top end mixture strength is not rich enough only the area between the middle and point of the needle will require reprofiling (the easiest part to reprofile). If the needles you choose prove too rich, at least you will know that needles slightly larger in diameter (leaner) are going to be required. **All** engines should idle using approximately the 2nd measuring point position of the needle.

In the first instance, the new needles should be tried with firm piston springs (as opposed to light piston springs) and standard weight oil should be used to fill the damper assembly (dashpot). The fitting of firm springs (8 grams, or heavier) will ensure that the piston is not able to rise too quickly and cause a 'flat spot.' At this point the only disadvantage of having fitted firm piston springs will be

	ABJ	ADS	ABN	ABP	AAY
1	0.089in	0.089in	0.089in	0.089in	0.089in
2	0.085in	0.085in	0.0855in	0.0855in	0.0855in
3	0.083in	0.0838in	0.0832in	0.0831in	0.083in
4	0.080in	0.0811in	0.0805in	0.0808in	0.0807in
5	0.0765in	0.0768in	0.0777in	0.0787in	0.0782in
6	0.0725in	0.0739in	0.0754in	0.0766in	0.0758in
7	0.0715in	0.0709in	0.0732in	0.0743in	0.0744in
8	0.0686in	0.0691in	0.0712in	0.0722in	0.0732in
9	0.0658in	0.0682in	0.0692in	0.070in	0.0723in
10	0.0647in	0.0662in	0.0672in	0.0685in	0.0717in
11	0.0636in	0.065in	0.0652in	0.0672in	0.0712in
12	0.0625in	0.064in	0.0632in	0.0659in	0.0707in
13	0.0614in	0.0627in	0.0612in	0.0646in	0.070in
14	0.0605in	0.0616in	0.0592in	0.0633in	0.069in
15	0.0595in	0.0603in	0.0572in	0.062in	0.068in
16	0.0584in	0.059in	0.0552in	0.0607in	0.067in

Chart shows five example needles from the *SU Reference Catalogue* suitable for a single 1 1/2in SU fitted to a 1100cc engine. Only the needle measuring points within the working range of the needles need to be recorded (to the 11th measuring point in this instance), and these will allow a direct needle comparison. It is exceedingly likely that the richest needle from the 2nd to the 5th or 6th measuring point (in this case ABJ) will run any 1100cc engine reasonably successfully.

that the engine will not accelerate as well as it might if it were fitted with lighter piston springs. However, getting the engine to perform without any hesitation or 'flat spots' from off idle through to high rpm, even if the acceleration rate of the engine is a little slow, is the first objective.

It is worth noting that the piston spring offers adjustable (via alternative springs) resistance to the rising piston. The stronger the spring, the lower the piston will 'ride' in the carburettor body when the engine is running at low rpm. A lighter spring will have the opposite effect. This is quite a significant feature with regard to engine economy but there is a limit to its usefulness. That limit is engine response. The piston spring can only be so strong before engine acceleration and overall engine performance are impaired and, at that point, the quest for fuel economy becomes self-defeating. As a general rule, fitting a light piston spring (2.5 or 4 grams) and a suitable needle, or needles, will result in good engine performance.

The oil in the damper assembly (or "dashpot" as it is commonly known) has the most effect at low rpm. The effect of the oil in the damper assembly is to dampen the what could otherwise be rapid upward movement of the piston. The damper mechanism smoothes out the piston movement and, especially at low rpm, prevents the piston lifting quickly and leaning out the mixture which would cause the engine to hesitate badly. The damper mechanism is, in fact, not really necessary at high rpm. The spring and oil dampening system works well on standard engines. With no oil in the dashpot a standard engine seldom runs well at low speeds and, if the piston spring is left out, a standard engine will not run well at all.

Make a note of each change made to the needle, or needles, between tests. This gives you a very clear picture of what has been achieved, what the effects of the changes are and where you have come from. It is really not difficult to get the needles 'right' for any application if you approach the task methodically. If needles are too small in diameter (rich) there is no alternative but to replace them with alternative needles of larger proportions.

Chapter 7
Carburettor tuning sequence

Irrespective of whether your engine is standard or modified, it still needs to be tuned correctly. The information in this chapter applies to any engine. The sequence is logical. There is a proviso on the condition of the carburettor, or carburettors, being tuned: no carburettor that has a mechanical defect of any sort, including leaks, can be tuned.

The first things that must be checked on any carburettor are the condition of the float, the needle valve's needle and seat and whether the shut off height (float height) is correct. If there is any doubt as to the integrity of the float and needle valve system, identify the carburettor model and order the correct new parts. After fitting the new parts check that the fuel level shut off height is to specification. Aim for a shut off setting in the middle of the tolerance range. Use thin fibre washers, thin aluminium washers or paper (fuel resistant) washers to pack out the seat if necessary. Details of all SU carburettor standard settings are available from Burlen Fuel Systems.

Warning! - A fuel level that is too high will result in a flooding carburettor (the fuel is able to dribble out of the top of the main jet).

The needle must be positioned in the piston correctly (shoulder of the needle flush with the piston surface).

Note that no engine can be succesfully tuned if there is a mechanical problem, or if it is excessively worn. Electrical systems must be in perfect working order and the ignition timing should be appropriate for the engine's state of modification.

BASIC/START-UP SETTINGS

Set the main jet, or jets, down from the top of the bridge/s as follows:
1 1/4in SUs: 0.030in/0.075mm
1 1/2in SUs: 0.040inin/1.0mm
1 3/4in SUs: 0.050in/1.25mm
1 7/8in & 2in SUs: 0.060in/1.5mm

These settings will ensure that the engine will start even though the settings will not necessarily be correct. Slight adjustment will almost always be required from this initial position to effect correct idle mixture.

The idle mixture of the SU carburettor is controlled by the setting of the main jet height (down from the top of the bridge). Virtually no engine

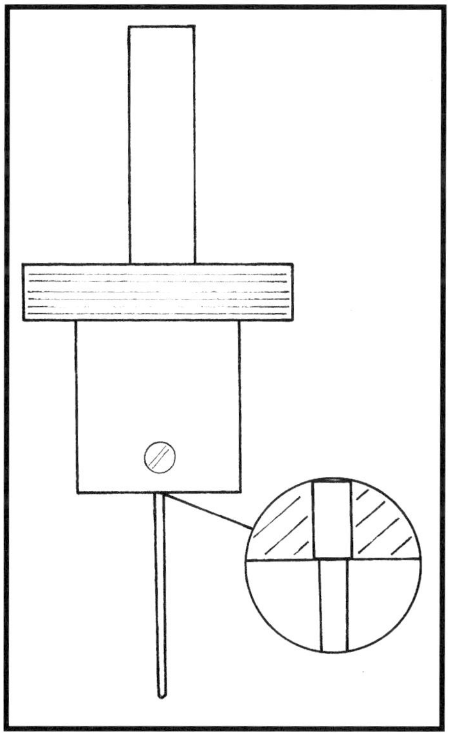

Needle correctly positioned in piston. Shoulder (1st measuring point) flush with the piston surface.

will run with the main jet flush with the top of the bridge: the rare exceptions feature a needle which is smaller than normal in diameter in the 1st and 2nd measuring point positions.

The needles that will require a different main jet depth setting are those that are approximately 0.003in/ 0.075mm to 0.004in/0.1mm smaller than the main jet diameter at the first measuring point, but there are not many of this type and such needles always supply a rich mixture for the size of carburettor. The exhaust gas COs of such needles will be high (at least 5% CO) at idle, which is too much for a road going engine. In the majority of instances (where a needle is 0.001in/0.025mm smaller than the main jet diameter) the exhaust COs of the engine will be between 1.75% and 2.5%. Just because an engine is a performance orientated engine, or even a pure racing engine, it does not mean that the mixture has to be super rich at idle. Engines go better when the air fuel mixture is 'right' for the particular engine. Engines that simply have a rich mixture all through the rpm range do not go as well and also consume huge amounts of fuel, most of which is not burnt at all. The correct full power mixture (an air to fuel ratio of 12:1) for engines gives 5-5.5% CO.

The majority of needles are 0.001in/0.025mm smaller than the main jet diameter at the first measuring point and approximately 0.003-0.005in/0.075-0.125mm smaller in diameter at the second measuring point.

It's easy to check that the main jets are all set to the same heights on any multiple SU carburettor set-up. The check involves removing the vacuum chambers from the carburettors to allow the height of the main jet to be measured with the depth tail of a vernier calliper. There is

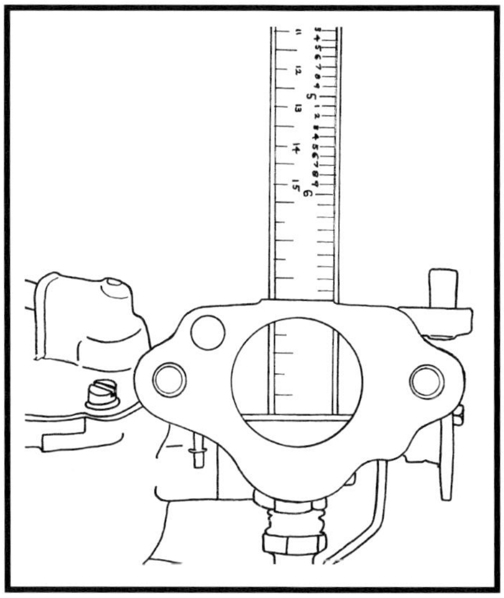

The tail of a vernier gauge is used to measure the distance from the bridge of the carburettor body to the top of the main jet.

little room for error with direct measurement.

The optimum position for the main jet/s can only be determined once the engine has been run. Using the start-up setting given here may mean that the idle mixture is too rich/ lean (CO reading) when checked with an exhaust gas analyser. Essentially, the mixture is set for the smoothest idle.

With 'new' multiple carburettor set-ups, or those where settings have been lost, it can be difficult to get airflow equal enough to start the engine and get it to idle well. In the first instance, to get the butterflies near equal, the throttle bar clamps are undone so that each carburettor's idle adjustment screw can be independently adjusted. Undo each idle speed adjusting screw (turn the screw anti-clockwise) until the spindle is seen to stop moving and then turn the idle speed adjustment screw inward until the spindle is seen to just move. Turn the idle speed adjustment screw a further half turn inward/ clockwise (use the slot of the adjusting screw as a visual guide and simply turn it through 180 degrees). Retighten the clamps in the linkage. The engine can now be started with the knowledge that the carburettors have a similar amount of butterfly opening set.

Note that the springs fitted under the idle speed adjustment screws will often be weak on well used or older carburettors and this can be the cause of the idle 'going off' only a short time after being set (screws move). It's advisable to replace these springs with strong new ones to prevent the screws from losing their settings through vibration. Sometimes the simple act of putting a washer underneath the spring will be enough to increase or restore the spring tension.

Important note - If your application features a distributor with a vacuum advance system, the system **must** be disconnected and the vacuum take-off in carburettor or manifold temporarily blocked.

IDLE SPEED SETTINGS

On multiple carburettor engines the throttle linkage between carburettors will have to be connected to start the engine but, once the engine is warm and idling (however roughly), the clamping arrangement that links the carburettors together can be undone and each carburettor adjusted individually to equalise airflow. The days of hose pipes and listening to the 'hiss' are long gone: a calibrated meter is what is required, and nothing less will do. These devices are made by several companies and are readily available at reasonable cost (Gunson, for example). On all SU carburettor set-ups the individual carburettor idle adjustment screws are what is used to synchronise the airflows of the individual carburettors at idle speed.

The air cleaners will have to be removed to synchronise the carburettors, because most airflow meters are designed to fit over the carburettor mouth.

The principle objective is to get the engine to idle anywhere from 800rpm to 1500rpm (depending on the engine modifications - the longer the camshaft duration the higher idle rpm), with equal amounts of air going through each carburettor and with 1.75%-2.5% CO for a road going engine and 3.5%-4.5% CO for a pure racing engine. The percentage of CO is measured using an exhaust gas analyser of which there are several examples on the market, such as Gunson's to name but one. The main jet is raised or lowered (in one flat of a hexagon [quarter turn of the adjusting screw - HIF] increments) to obtain the smoothest idle at a realistic idle speed for the particular engine and, to a certain extent, the particular needle fitted to the carburettor.

Note that it is quite possible to get the idle exhaust emissions down to 1% CO, or less, with an SU but having a 2% CO reading is very acceptable and often easier to attain. Virtually any SU can be set to idle with a 1% CO (or less) reading, but the instant the engine is required to work 'off idle' the air/fuel mixture will usually be too lean and the engine will usually hesitate or stall. Very few needles are available that alter in diameter (become smaller) quickly enough to offer suitable air fuel mixture ratios conducive to good engine performance once off idle. You can however 'make' a needle that will allow very low idle COs and yet pick-up instantly. The diameter from the 2nd to 3rd needle measuring points will have to alter sharply (get smaller quickly). It is even possible to get an engine equipped with SUs to idle with 0.3%-0.5% CO and still have good engine pick-up off idle, but you can't buy a needle to do this; you'll definitely have to 'make' it. Most people do not bother to go to these extremes, but it is definitely possible. The needle is stepped so that at idle the mixture is very lean. However, the instant the needle lifts, it presents a smaller diameter and supplies the correct mixture from off idle to full revs.

In most instances for road use a 2% CO reading will be accepted as good and in most countries is within the legal limits for a non-catalytic converter engine. Catalytic converter engines do not show low CO because they are set so lean; rather they show a zero to 0.5% CO reading because of the catalytic converter (if the 'cat' was removed most such engines would have between 2 and 2.5% CO at idle).

An engine used only for racing should still be set at no more than 5% CO as there is no point in being wasteful and, further to this, the engine will run better with a leaner mixture and there is less likelihood of plug 'fouling'.

So, the acceptable range of idle speed exhaust gas CO emissions is from 2% CO on the lean side to 5% CO on the rich side. The recommendation is to set the fuel/air mix as lean as possible within the given parameters and without the engine being prone to 'running on' when the ignition is switched off, or having hot starting problems (too lean a mixture in each case).

It doesn't matter what type of engine you have or how many carburettors, the ideal is the same: reasonable idle speed commensurate with the type of camshafts fitted to the engine and a 2-5% CO reading with all carburettors flowing the exact same amount of air.

Changes can then made to the amount of fuel being admitted to the engine by raising or lowering the main jet. The main jet is lowered to richen the mixture and raised to lean it. To make an adjustment the pistons are removed and the main jets raised or lowered (settings measured and equalised using vernier measurement). With the needles positioned correctly in the pistons, the airflow equalised and the distance from the top of the main jet to the top of the bridge known, the amount of fuel being admitted into the engine through each carburettor will be equal. Note that the reason for using direct measurement of the distance between the bridge and main jet is that it is otherwise quite easy to 'get lost' with turns and part turns of adjusting nuts and so on.

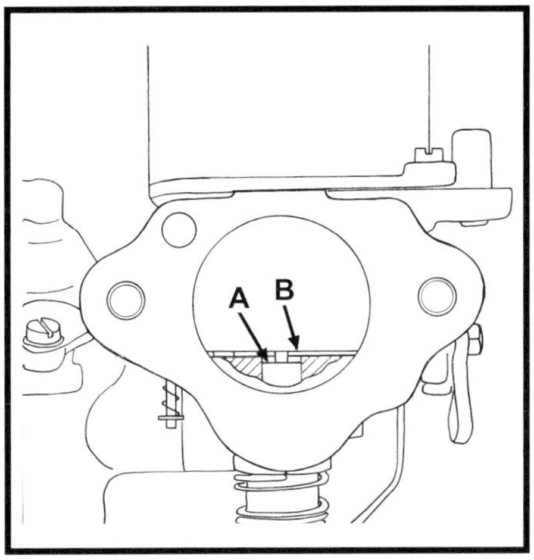

The cutaway section of this diagram illustrates a main jet (A) which is well below the top of the bridge (B).

The final setting of the main jets is the one which sees the engine idling smoothest and with acceptable COs (1.75% minimum to about 3.5% CO). Irrespective of where the main jet height is set to effect correct idle speed mixture, the distance down from the bridges to the top of the main jets in *all* the carburettors being used must be the same (within 0.002-0.004in/0.05-0.1mm of each other) as measured with a vernier calliper. If it is not possible to effect a good idle with good CO readings by having all the main jets at the same height there is something wrong, such as one carburettor having a worn spindle, main jet bore or needle.

It is not all that uncommon to see the main jets adjusted down from the bridge by as much as 0.080-0.090in/2.0-2.25mm on some of the larger SU carburettors. However, for the full range of SU carburettors expect the total range of acceptable adjustment to be from 0.020in/0.5mm to 0.090in/2.25mm.

Because of the accuracy of SU componentry and the carburettor as a whole, measuring down from the bridge of the carburettor to the top of the main jet is an ideal method of setting SU carburettor idle mixture and effectively 'knowing where you are.'

Expect a standard engine to idle at 500-700rpm. Expect a modified engine with a 270 degree high performance camshaft to idle at approximately 700-1000rpm. Expect an engine equipped with a 280-290 degree duration camshaft to idle at 900-1200rpm and any racing engine equipped with a 300 or more degree camshaft to idle at approximately 1200-1500rpm.

At this point you should have an engine running as smoothly as possible with a good CO reading and with perfectly equal airflow through each carburettor.

Idle speed advance

If your engine does not seem to respond to main jet adjustment, check to see how much idle speed spark advance the engine has. This is a **vitally important** part of engine tuning which is often missed. No engine can run correctly without the right amount of spark advance. Expect a mildly cammed engine (270 degree cam) to need 10-12 degrees of idle speed advance and a medium cammed engine (280-290 degree cam) to require 12-16 degrees of idle speed spark advance. Some racing engines equipped with 300 degree duration camshafts may well require idle speed advance amounts of up to 20 degrees, but usually 16 degrees will prove sufficient.

Essentially, any engine needs to be set with an amount of spark advance that allows it to attain the smoothest idle possible. Check the idle speed advance using a strobe light and consider up to 20 degrees as not being out of the way if the engine responds to it. Any engine can be safely tested to see what idle speed advance it responds to as there is no load on the engine at idling speed. **Caution!** - If your engine responds to having a lot of idle speed advance, the ignition system will have to be modified; otherwise it'll end up with too much total advance which could cause engine damage or lack of top end power, or both.

A sensible amount of idle speed advance that allows the engine to attain the smoothest possible idle is what is required. Engines that do not have enough idle speed advance frequently 'spit back' while idling (backfire through the carburettor) and hesitate when initially accelerated, but then pick up and go. It would not matter how rich the mixture was made, the engine would still spit back because the problem is not being caused by air/fuel mixture but by the fact that the spark is timed too late.

Vacuum advance

When the vacuum advance (if applicable) is reconnected after carburettor tuning the engine might develop a light misfire. If it does, check to see how much advance there is with the vacuum advance system working. If the extra advance is more than 10 degrees then this is likely to causing the misfire. It's not impossible for engines to have as much as 20 degrees of extra advance at idle, and above, when revved in unloaded condition.

If the engine runs okay under normal load, leave well alone. However, if it misses on a very light loading the solution is to limit the amount of vacuum advance possible (cut it back from 20 to 15, for example). This can be done by replacing the vacuum advance canister with one that has less advance built into it or by restricting the amount of vacuum that can get to the vacuum canister.

SYNCHRONISING MULTIPLE CARBURETTORS FOR OFF IDLE USE

With the engine able to idle correctly the next step is to synchronise the carburettor butterflies for off idle work (that is any amount of butterfly movement off the carburettor idle adjustment screw stops). With the engine warmed-up, idling correctly and with the air cleaners removed, loosen all side by side throttle bar clamps and any other linkage between the throttle bar and the main throttle arm that is directly controlled by the accelerator pedal. Different applications will have different set-ups to operate the carburettor butterflies, but all will have an arm directly controlled by the throttle pedal, and a rotating bar with links to each carburettor's butterfly; the arm will also feature some form of clamping arrangment to allow throttle (butterfly) synchronisation.

Note that some assistance will be necessary; it is just not practicable to do all of the work single-handed.

Warning! - When working on a running engine make sure you aren't wearing **loose clothing** and be fully aware of the danger that moving components and hot components represent. Be **very careful** where you put tools.

In the first instance, with the engine running at idle, tighten the clamps on the throttle bar which link the carburettors on the same side of the engine. When all clamps are fully tightened check the airflow of each carburettor. The readings should all be the same as they were when the clamps were loose. If they are not, undo the clamps, retighten them and then check the airflows again. Repeat this procedure until the readings are all equal.

Now have an assistant apply a little throttle: the place to apply light hand pressure is on the actual throttle arm (almost always between two carburettors), not on one end of one of the carburettor spindles or on the accelerator pedal. This method reduces to a minimum spindle and throttle bar flexing which could lead to one carburettor being given slightly more throttle. Check the airflow reading of the carburettors once again. The airflow must read the same for each carburettor (the reading will have changed now that more air is being drawn in) regardless of how many carburettors there are. When the individual carburettor spindles are clamped together, the carburettors effectively become a single unit so that what happens to one happens to all.

If the engine is of Vee configuration, for example, fitted with four carburettors, clamp the carburettors on one side of the engine and put a small amount of movement into the throttle bar and check the airflow readings of those carburettors. They must be equal. Do this on both sides of the engine. At this point the carburettors on each side of the engine are synchronised with each other but the two banks of carburettors are not connected or synchronised. Synchronising the two throttle bars is the next step.

Most throttle linkages tend to have a degree of 'slop' in them, and the older they get the more slop they tend to have. All obviously worn parts, such as bushes and joints, for example, should be renewed.

Next, two people need to be involved. With the engine switched off an assistant will be required to put a small amount of movement into the accelerator pedal and hold it steady.

With light finger pressure, push the throttle arm back (in the opposite direction that it moves to open the throttle). Tighten the clamping screw that secures the throttle arm to the throttle bar. By placing a small amount of tension on the accelerator pedal any remaining slop should all be in one direction and, to a degree, nullified. After the clamping screw has been tightened your assistant can remove his/her foot from the accelerator pedal.

With all of the clamps tight the engine can be started. With the engine idling the airflow of each carburettor can be checked to make sure that the readings are exactly the same for *all* carburettors and that the readings are the same as before when the throttle linkage was not connected.

The next step is to have your assistant depress the accelerator pedal and take a reading once again. The airflow readings should be identical for each carburettor for any given amount of throttle bar movement.

For Vee engines, for example, the two banks of carburettors must be linked so that when the accelerator pedal is depressed, by any amount, all carburettors have the same amount of airflow. This is reasonably easy to check, but the throttle bars must remain stationary while the airflow meter is moved from one side of the engine to the other. If the two banks of carburettors are not synchronised and flowing air equally, performance will be down as one bank of cylinders will be receiving more air than the other. In this situation having two flow meters is ideal. The two flow tubes of two Gunson Carbalancers can be taped together and viewed at the same time.

The throttle linkages on some multiple SU set-ups can be unbelievably complicated. For example, the twin SUs on a Daimler SP250 V8 have a very complicated linkage which is difficult to set up. In these worst case scenarios, what is important is to have the arms that pivot set so that the arc of travel is equally dispersed about the horizontal plane. Also the linkage arms must be set so that they have the same centre-to-centre lengths (check off engine with a vernier caliper).

SUMMARY

With carburettor airflow synchronised at idle and off idle, and the main jets set to equal depths below the bridges to effect correct idle air/fuel mixture, the only further adjustment required is to the off idle mixture via replacement or modified needles.

Chapter 8
Rejetting (engines equipped with SUs as standard)

This chapter is for cars with a high-performance engine that was fitted with SUs as original equipment. Engines like this are usually quite mildly tuned in their standard form, those used in MG Midgets, Mini-Coopers and MGBs are good examples. Engines like this are often being slightly modified when rebuilt, while overall retaining the majority of their original equipment specification. Often, after a rebuild, especially with minor modifications, the original jetting is no longer quite right.

The alterations necessary to rectify this sort of situation are usually quite minor, and always confined to the needles and, perhaps, the piston springs (replacement of a spring which no longer has the standard rated tension). This means that new piston springs may well need to be fitted and the existing needles altered or alternative needles substituted. It's often expedient to simply alter the original needles. For a slightly modified engine the standard needles will never be all that far from correct, mid-range richness (that's the 3rd to 7th measuring points) usually being the problem.

If the engine you have is a slightly modified version of a standard factory performance engine (twin carburettor engine for instance), the chances are that the standard jetting will work well enough but not be totally suitable. In this situation, try the rich needles that are usually listed as alternatives. Even the richer needles available from the SU range, which were fitted to the particular engine type at one time or another, may not prove to be correct. Try the factory listed rich needles and recommended piston springs by all means, but they will not always be right if the engine is very far from standard specification, so be prepared to swap or alter needles as necessary. The recommendation is to stay with the standard needles and piston springs and modify the needles in the first instance.

Important - The following procedure assumes that the carburettors are of the type and specification that were factory fitted to the particular engine (not larger carburettors) and are in totally serviceable condition; the engine is mechanically sound; the electrical system is sound; the ignition timing correct and the throttle linkage in good serviceable condition.

Note that having the right amount of ignition timing is vital if the best possible engine performance and economy is going to be realised. Retarded engines (the usual scenario) are not efficient air/fuel burners.

Usually people just switch the original needles to known richer needles, or needles that someone tells them might work, and hope that the needles will in fact be right for the application. This is simply taking a 'stab in the dark' and is generally a waste of time, often little or no real progress is made: there is a better way.

Standard piston springs are rated at certain compressed heights (listed on page 158 of the *SU Reference Catalogue*). Essentially the A-type Light Blue is 2.5oz, Red 4.5oz, Yellow 8oz and Green 12oz, and they are used in 1 1/4in, 1 1/2in and 1 3/4in SUs. Other A-type springs are Red

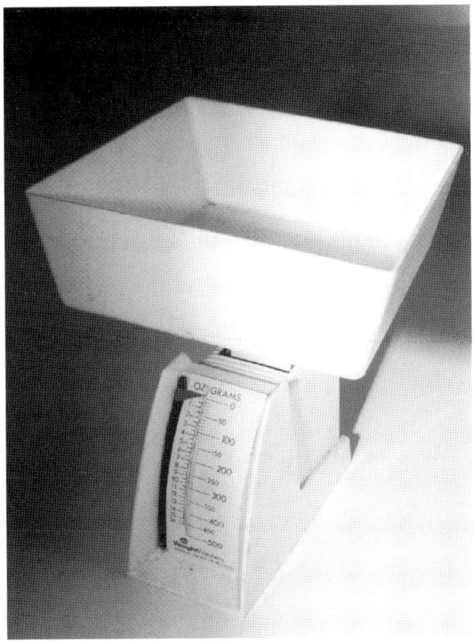

Inexpensive scales are available which can easily measure to the required accuracy.

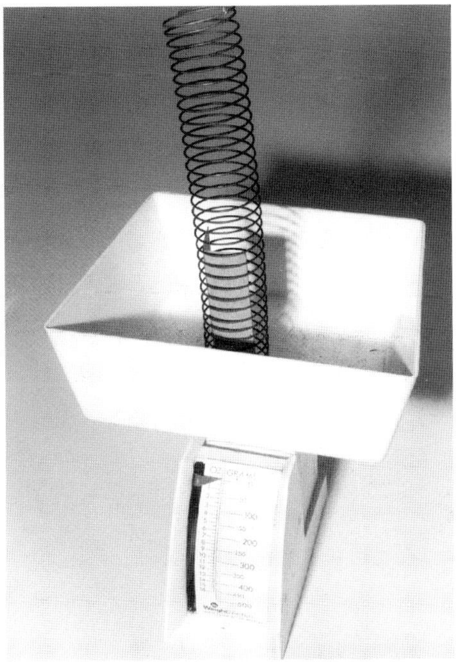

Cut a piece of card to compressed height of the type of spring being tested. Fold the card so it fits inside the spring. Then compress the spring by hand.

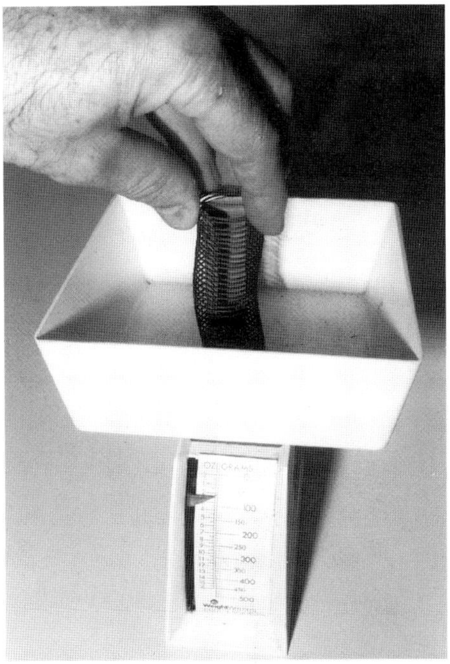

With the spring compressed to the rated height, the reading is taken off the scale. The bent card also keeps the spring in line.

and Green (11.25oz), the Light Blue and Black (4.5oz) and the Light Blue and Red (18oz). B-type springs are used in HIF carburettors of all sizes. The Red B-type spring is 4.5oz, Yellow 8oz, and Green 12oz. There is no 2.5oz piston spring for a 2in SU or any 2.5oz B-type piston spring. Many enines have 2.5-4.5oz springs fitted as standard.

Better engine performance is almost always available if light piston springs are used (2.5 or 4.5oz, for example) and the needle sized to suit the spring.

The paint applied to the piston springs by the factory when they colour coded them often disappears after the springs have been cleaned a few times with solvent. This can present a problem in identifying exactly which spring(s) you actually have. If you have any doubt about the springs you have, the easiest solution is to buy new ones of the correct original specification.

Another solution to this problem is to measure the spring(s) you have and re-categorise them. This is not a major problem at all, and is achieved by checking the pressure in ounces of the spring you have using a small set of calibrated scales. Scales that are suitable for this sort of checking operation are not expensive. These types of scales only go up to 16oz, but this is enough. The spring is rested on the scales' tray and the scale reset to zero. The spring is then compressed by hand to the required height as indicated by a piece of marked card.

It is reasonably easy to distinguish which spring is which. Used springs do not always make the factory recommended pressure at the listed height (anything up to 1oz less than rated). What is really important, however, is that on multiple carburettors the spring pressures of all the springs be equal at the factory rated height. What matters is that each of the springs is exerting as near as possible the same amount of resistance to the rising piston of each carburettor: if the springs are different, there will be slight mixture differences and this is not correct.

Having two worn piston springs in a pair of carburettors (lighter than standard but of equal pressure) is, however, often quite acceptable. The necessity for equal spring pressure is often not taken into consideration when there is more than one SU on an engine and problems which are not easy to track down may result. The principle of tuning SUs for any engine revolves around first setting the piston spring pressure and then finding needles to suit the piston spring.

The next part of the process involves removing the damper and fitting a calibrated rod into the top of the carburettor (oil in the dashpot

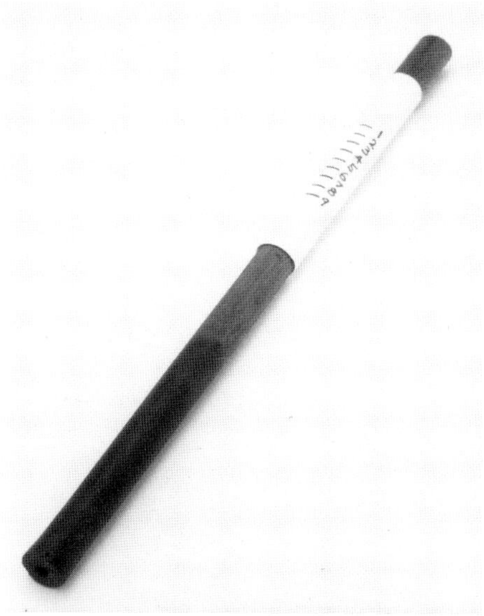

This calibrated rod is made of a 6 inch length of bamboo. A stick-on label has been wrapped around it and the measuring point lines drawn on to it.

removed). You cannot buy these rods and so will have to make them. This is not a difficult task as all that is required is a lightweight rod of some description (*e.g.* wooden dowelling, aluminium or plastic tubing) that can be accurately marked with the SU needle measuring point positions. Each carburettor must have its own calibrated rod so that it can be clearly seen that each carburettor piston is lifting an identical amount for a given amount of throttle. The calibrated rods are required so that you can visually monitor the position of the rising piston and, as a consequence, needle position. The air/fuel mixture that is admitted to the engine is controlled by the height of the piston above the bridge of the carburettor body. The throttle (butterfly) can be fully open, but the air/fuel mixture is still controlled by the position of the piston.

The calibrated rods which should be of 0.312in/7.93mm diameter and 5-6in/127-152mm long are marked with rings 1/8in apart which represent the needle measuring points.

Engine not running: the piston is resting on the bridge and the main jet is wound down from the bridge. The first measuring point is, as a consequence, above the top of the suction chamber.

The first needle measuring point is, in practice, made higher than the top of the carburettor's vacuum chamber to take into account the distance between the main jet and the top of the bridge, and the distance the stop is set proud of the base of the piston. These measurements are taken *after* the engine's idle has been set and jet heights set to optimum. With these two factors taken into account, the measuring point positions on the calibrated rods will be dead accurate. So, when the engine is not running, the 1st measuring position will be above the top of the vacuum chamber by around 0.020in/0.5mm to possibly 0.060in/1.5mm, or even more, depending on the main jet settings. When the engine is idling, expect the 2nd measuring point of the needle to be at least level with the top of the vacuum chamber. Idle usually corresponds to the 2nd measuring point, though higher and lower variations are sometimes seen.

Engine idling: the reading now shows the 2nd measuring point. This is a low reading, most engines will give a reading between the 2nd and 3rd measuring points.

With the standard recommended needles fitted, the engine can be started. The first and second measuring points will be suitable for starting the engine and being able to set the idle.

When the engine is off idle, the piston will have risen from its idle position, as will the calibrated rod (they are directly connected). As the piston lifts the markings on the calibrated rod will be clearly seen above the top of the vacuum chamber and the number of marks can be counted to find which measuring point is lean. When the engine hesitates the mixture is lean. In the first instance the throttle is opened very slowly until the engine hesitates. This will usually happen somewhere around the 3rd or 4th, perhaps even the 5th, needle measuring point position. The measuring point at which the engine hesitates is noted,and the engine turned off; the needles are then

Piston, and as a consequence, the calibrated rod, rise to the point of hesitation (lean). Polish the needle at the measuring point indicated by the calibrated rod.

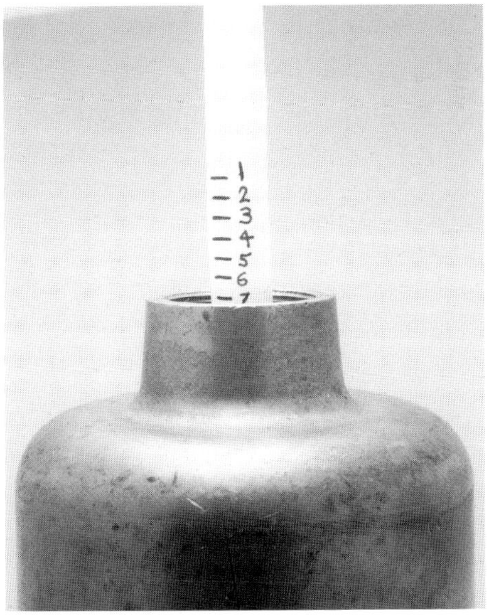

Piston, and as a consequence, the calibrated rod, lifted to the 7th measuring point without any hesitation. Needles need to be lean initially and made richer.

removed from the carburettors and polished down in the exact position indicated using the method described in chapter 5. If the noted position is midway between two measuring points, polish the needle there.

Needle polishing is the easiest and quickest way to achieve the correct mixture. If new needles of different specification are going to be bought it is quite likely that the right needles can be procured from an SU agent, but this will take time. Resorting to polishing the needle is a very practical option which, if done carefully, gives an excellent result and, of course, once the ideal needle specification is known you can purchase matching new needles with confidence.

The testing and polishing procedure should be continued until the engine no longer hesitates at that particular measuring position. Repeat the process as necessary at different measuring points (and progressively higher revs) until the engine does not hesitate up to half throttle. The reason for only going to half throttle is that mixture problems usually occur before this point and it is not good, or necessary, to rev the engine excessively while it is not under load. Once an engine is able to go to the half throttle position (using very slow throttle opening) without hesitation, the correct fuel mixture beyond that is not too difficult to arrive at.

Ideally, there should always be a 0.001in difference in needle diameter from one measuring point to the next for most of the needle's length (some needles do become parallel as they near the point - usually the last 4 measuring points).

The initial object of this exercise is to be able to open the throttle slowly (and it does not matter how slowly) up to about half throttle without the engine hesitating at all. Half throttle equates, approximately, to the 5th or 6th needle measuring points on 1 1/4in SUs, the 6th or 7th on 1 1/2in SUs, the 7th or 8th on 1 3/4in SUs and the 8th or 9th on 2in SUs. There must be **no hesitation whatsoever** as the rpm is increased. If necessary, keep polishing the needle at the relevant measuring point until the mixture is rich enough at all points up to half throttle. The calibrated rods are ideal for showing the exact point of leanness during progressive throttle opening.

There is no guesswork necessary in altering needles. Most needles will only need 4 or 5 needle measuring points altered and, often, as little as 0.0005-0.001in to a maximum of 0.002in of diameter removed at the individual measuring points.

With the damper rods still out of the carburettors expect the engine to hesitate when the throttle is opened *quickly* to the half throttle position. Even though this second phase is essentially a repeat of the initial procedure, it's necessary to do it this way around. By sizing the needles to a very slow throttle opening rate, a baseline sizing is arrived at for mixture strength. As a consequence of doing this, the basic needle proportions have been finalised. It is easier to start with the mixture lean and polish the needle bit by bit until it's right, than to polish the needle a large amount and end up making it very rich: this is not the way to best power.

The individual measuring point diameter sizes of the needles will perhaps have to be changed further to give a richer mixture, but the overall shape of the needle is now set. Some of the individual needle measuring point sizes may need to be changed by between 0.001-0.002in, or even more in some cases, to give more mixture strength to suit a quickly opened throttle.

The throttle is not 'snapped open,' as in being instantaneously opened from idle to just past half throttle, but opened reasonably quickly (half a second). This could be a quick wrist movement on the actual throttle bar or an accelerator pedal movement. The throttle is opened to half throttle and then released **immediately**. In the time it takes to open the throttle in this manner, the engine revs must rise without any hesitation whatsoever. The slightest hesitation will show up when the engine is under load as a 'flat spot.'

If the throttle is opened very quickly, as in being 'snapped open,' the engine will almost certainly hesitate, especially low down at the 3rd, 4th, 5th and possibly the 6th needle measuring positions. Once the needles have been sized for quick throttle opening to the half throttle point, any throttle action past half throttle (or above the 6th or 7th measuring point position) will not normally cause the engine to hesitate, even if the throttle is moved very, very quickly from, say, half to three-quarter throttle. **Caution!** - opening the throttle in this manner does no real harm to an unloaded engine, but persistent free-revving of any engine is **not** recommended and it is not necessary in order to tune SUs.

It is quite usual for the 3rd, 4th, 5th, 6th and 7th measuring point sizes to all require size changes to allow clean running under quick throttle opening. This is because the engine is being accelerated and there simply has to be sufficient fuel available to match the increased air intake.

If, for example, the needle requires resizing between the 3rd measuring point position and the 8th measuring point position, the 9th, 10th, 11th, 12th and 13th must each be smaller, even if only by 0.001in (0.0005in for some smaller capacity engines) than the measuring point size that immediately precedes it. Note that small engines (1000cc, for example) with twin carburettors that are a little too large may respond to a 0.0005in reduction in needle point diameter, but usually 0.001in will make a difference. This means that all of the measuring point sizes after the one being worked on are always progressively reduced in size as the needle is being reprofiled: this ensures the needle is always getting slightly richer as it rises through the main jet.

Once the damper is fitted back on a carburettor, the damping mechanism will be in operation and there will be resistance to the piston rising quickly. This factor is what stabilises mixture changes. The piston will, of course, lift and lift right to the top against the spring pressure, but it will never lift quicker than the damper mechanism will allow (unless there is no oil in the dashpot). The amount of resistance offered by the damper mechanism is virtually always enough to slow the rate of rise of the piston enough to allow clean engine acceleration under load.

This is where needle diagrams are so handy because all of the needle sizes that have been used can be recorded and referred back to when necessary. There are two pages of blank needle diagrams at the end of this book which can be used to record the needle sizes you use to narrow down the needle measuring points that suit your particular engine. Each needle is a development of the previous needle.

SUMMARY

This chapter has dealt with re-tuning an existing engine which has been slightly modified from standard. The carburettor parts can all be as per original equipment with, perhaps, the exception of the needles. The needles can either be modified originals or alternative specification needles. If the original needle is lean in a particular area (found by using the calibrated rod method) an alternative needle can be found by looking through the *SU Carburettor Needle Profile Chart* book. Select a suitable needle which has a smaller needle diameter at the particular measuring point. This will richen the mixture in that area. Needle polishing is recommended as an expedient way of ending up with the 'right' needle for the application very quickly.

A good example of using the whole range of standard needles as a guide to choosing a needle or modifying a needle to suit a modified version of the standard engine is the Rover V8. The list of standard needles used over the years is: KL, KO (early engines), BAC, BBG, BBV, BAK, BCP, BBU and BFW.

The BAC has a very rich top end mixture as indicated by the measuring point sizes which start to change dramatically from the 7th measuring point down. This is a heavy duty application needle and the top end mixture has been made rich to suit the probable engine loading one of these vehicles (Land Rover) could have to contend with. Don't expect to have to use more top end richness than this for any application.

All the needles follow a similar pattern from the 1st to the 7th measuring point size. The average top end mixture measuring point size is about 0.069in. The KO needle is a good 'middle of the road' needle from this selection and, therefore, an ideal starting point needle. The problem for later SUs is that this needle is not available in spring loaded (swinging needle) form: however, a needle could be altered to the KO sizes.

Rover V8 standard needles

	KO	KL	BAC	BBG	BBV	BAK	BCP	BBU	BFW
1	0.099	0.099	0.099	0.099	0.099	0.099	0.099	0.099	0.099
2	0.095	0.095	0.095	0.095	0.095	0.095	0.095	0.095	0.0962
3	0.0925	0.0925	0.0932	0.0932	0.0932	0.0932	0.0929	0.0932	0.0939
4	0.0895	0.0895	0.0907	0.0905	0.0907	0.0907	0.0905	0.0907	0.0914
5	0.087	0.087	0.0875	0.0878	0.0875	0.0875	0.0881	0.0876	0.0887
6	0.085	0.085	0.0852	0.0852	0.0852	0.0852	0.0858	0.0859	0.086
7	0.0825	0.0832	0.0823	0.0829	0.0829	0.0823	0.0835	0.084	0.083
8	0.0795	0.0792	0.0763	0.0806	0.0805	0.0792	0.0804	0.0822	0.0799
9	0.0765	0.076	0.0703	0.0783	0.0773	0.076	0.0793	0.0805	0.0768
10	0.074	0.0729	0.0642	0.076	0.0742	0.0729	0.0781	0.0788	0.0736
11	0.0715	0.0697	0.058	0.0737	0.071	0.0697	0.0759	0.077	0.0705
12	0.069	0.0665	0.052	0.0713	0.0679	0.0665	0.0737	0.0752	0.0674

All of standard needles listed for the Rover V8 are similar up to the 7th measuring point size and none will be found to be too rich low down in the rpm range; if anything, they'll be slightly too lean. It doesn't always take much of a diameter change to a measuring point to make a difference to the mixture which changes an engine from one that is not running all that well to one that is running very well indeed. In practise, none of the later standard needles are really rich enough between the 2nd and 7th measuring points for a modified Rover V8 engine fitted with the standard twin SU and inlet manifolding system. However, the early KO and KL needles are about right.

The sort of spring loaded (swinging) needle that will give a richer mixture between the 1st and 7th measuring point positions is the BCW. This is a good starting point needle for a 3.5 litre Rover V8, but the top end mixture is usually on the lean side (even though needles with this sort of sizing between the 9th an 12th measuring point sizes were used as standard). The BCW needle can be altered so that it has more top end richness (that's from the 7th measuring point down). The BBW will often prove to have too much top end richness (depending on the engine modifications); while the low and mid range mixture strength (1st to the 7th measuring point) is about right as it's just a little bit richer than the BCW up to the 7th measuring point position.

Rover V8 'spring-loaded' standard needles

	BCW	BBW
1	0.099	0.099
2	0.095	0.095
3	0.0926	0.0923
4	0.0902	0.090
5	0.0879	0.087
6	0.0844	0.0832
7	0.0815	0.0792
8	0.0795	0.075
9	0.0775	0.0717
10	0.0755	0.0682
11	0.0735	0.0647
12	0.0716	0.061

The overall reason for having to 'make' needles is, of course, that no one standard needle suits a modified engines requirements. One standard needle may well be right for low end to mid range use, but be too rich at the top end: another needle may well be ideal at the top end but be too lean for low end and mid range use. At least, if two sets of needles give ideal performance levels in different rpm ranges, the parameters for an ideal needle are known and needles that represent a combination of the two needles can be fashioned.

Caution! - Running an engine lean can result in engine damage. An engine that is very lean will often backfire through the carburettor. Do not persist with an engine that you think is running lean. Make richer needles (0.001-0.002in smaller at the appropriate measuring point for the carburettor) and test the engine to see if the mixture is still lean. Sparkplug checks and scientific analysis (CO reading) are used to make sure that an engine is running with an optimum top end mixture.

An excessively rich mixture does not produce optimum engine power and is not conducive to the production of maximum engine acceleration. A rich mixture also leads to premature engine wear as the excess fuel is forever 'washing' the oil film off the bore walls. Little good comes from having a very rich fuel mixture except for when starting an engine, which is what the choke should be for. Having a permanently rich mixture is like running the engine with the choke partially out all of the time.

Choke

Be aware of the fact that when the choke knob is used and then pushed back in, the main jet (which drops to enrich the fuel mixture) **must** return to its original position. If this does not happen the idle mixture will be erratic. Worn linkages can cause this problem. Sometimes the solution can be as simple as fitting a return spring on the choke lever arm. The bottom line is that the main jet must return to its normal position after the choke is used

It's a good idea to lubricate the carburettor throttle spindles (and choke mechanism for that matter) occasionally using WD40 or CRC-type penetrating oils. Once every couple of months is about right.

Chapter 9
Rejetting (other engines)

Note that the line illustrations and photgraphs in the previous chapter are relevant to this chapter too.

The difference between this chapter and the previous one is that, for non-original SU applications, needles are likely to require attention to all measuring points except 1st and 2nd. This is because the engine will be unlisted in SU charts and, therefore, the needle/s you start with are likely to require more modification to make them perfect for your particular application.

Knowing what needle specification to buy or make is the crux of the matter and what causes most of the tuning problems related to SU carburettors. What usually happens is that a needle is found that produces a roughly correct mixture for around 60% of the engine's requirements. The engine goes well in many parts of its rpm range, but never in all of them. This leads to the engine having what is often termed 'a flat spot' which realistically means that the engine has a lean or rich condition (lean usually) at a particular point in the rpm range and does not accelerate cleanly from idle to full bore. This sort of situation leads to frustration with the SU carburettors and often their replacement with a different type of induction system which is thought to be more tuneable.

Such radical action is, in fact, quite unnecessary. Most carburettors require attention to detail to get the air/fuel mixture correct for the application. The SU is no different in this respect to any other carburettor, other than the fact that a needle (or needles) will have to be modified (physically or by substitution). This is a significant point, of course, as most other carburettors used for high performance engines can have parts interchanged to alter engine performance in about five minutes, but not always for the better - especially for the pocket. Tuning SUs is not like this at all: it takes time to determine ideal needle specification. That said, it doesn't take all that much time and only needs to be done once, unless the engine specification is subsequently altered.

The following procedure can be applied to *any* engine in *any* state of tune with *any* number of carburettors fitted to it. This applies whether the engine was originally fitted with SUs or a whole induction system has been specially made to allow the use of SUs. There is virtually no engine application that cannot be accommodated and the best possible engine performance obtained. The limit with each carburettor size is the fact that no carburettor can flow more air through it than its size allows. This ultimately means that if the carburettors fitted to the engine are too small, the engine performance may well be excellent - but only up to the point of maximum carburettor airflow as opposed to what the engine might be capable of flowing.

For the following procedures to work correctly the provisos are that the carburettors must not be too large (unrealistically large for the size of the engine on the basis of ccs per cylinder), the inlet manifold must be sized correctly for the cylinder head's inlet ports (slightly larger than the inlet port), the carburettors (if more than one) must be synchronised at idle and off idle, the carburettors must be in excellent condition and the carburettor butterflies must be fully open at full throttle. Any engine being tuned must be in excellent mechanical condition and have an electrical system that is in excellent order too. Poor piston ring sealing and/or leaking valves will have a profound effect on performance and will prevent optimal carburettor tuning.

PISTON SPRINGS

A very significant factor is that the piston spring/s must be chosen first. Piston springs come in different tensions and are carburettor model specific. The *SU Reference Catalogue* lists the standard range of springs and their respective tensions in ounces of pressure at certain compressed heights. The piston spring to be used is chosen *first*, then the needle is sized to suit the characteristics of using that spring in that carburettor on that engine.

The choice of piston spring tension will depend on how much performance is required and how important achieving maximum fuel economy is. Within reason, the stronger the piston spring the leaner the needle can be (to obtain best possible fuel economy) and, as a consequence, the slower the engine acceleration. Conversely, the weaker the piston spring the richer the needle will have to be; overall engine acceleration will be improved, but at a slight cost to fuel economy. The SU can in fact even be tuned to suit an individual driver's driving style. So, if economy is more important to you than having maximum engine performance, fit strong piston springs to your engine (8oz). If on the other hand all round engine performance at some small cost in economy suits you, fit light springs (2.5-4.5oz).

As a general rule a light piston spring is used in high performance applications (quick acceleration with a possible slight loss in economy). The fitting of the lightest spring conducive to good all round engine performance with economy is the ideal (usually 2.5oz or 4.5oz).

There is a range of ten springs available for SU carburettors and these are all listed in the *SU Reference Catalogue* on page 158, along with their respective tensions at certain compressed heights.

Note that when a piston spring is too strong for the application the engine will accelerate reasonably well, but as the revs get higher (three-quarters of the possible rpm with the particular engine), the engine will suddenly start to accelerate much faster. This is a clear indication of too much spring tension and a slightly weak mixture. Reducing the amount of spring tension is the very next step (8oz down to 4.5oz, for example).

If the piston springs you have no longer have any paint on them to identify them by colour code, they can be easily checked to see what their strength is. The *SU Reference Catalogue* gives gives pressure ratings for all spring types at a certain compressed height. Use a small set of scales calibrated in ounces and, with the scales zeroed out with the weight of the spring taken into account, press the spring down by hand to the required height. The spring will exert a force on the scales which can be read directly off the measuring scale. Any SU spring can be identified in this manner without guesswork.

IDLE SETTINGS

Important - If applicable, vacuum advance systems must be disconnected, and the vacuum take-off in carb or manifold temporarily blocked before following the idle setting procedure.

All engines equipped with SUs, irrespective of how many carburettors are on the engine and how many cylinders each carburettor has to feed, can be set to idle almost any SU needles. Virtually all SU needles which can be fitted into each model of carburettor have very, very similar needle diameters at the 1st and 2nd measuring points (often the 3rd measuring point too). This means that basically all SU needles can be used to allow starting and to effect the idle settings of any engine. It may take a bit of choke to get the engine going (use an 'engine start' spray, if necessary); once warm the engine's idle mixture will almost always be able to be finalised regardless of the overall needle specification.

The 1st and 2nd measuring points need only ever be the usual standard sizes as listed in the SU needle book. For example this means that for a 0.090in diameter needle the 1st measuring point diameter will be 0.089in and the 2nd measuring point 0.085in on average. A 0.100in diameter needle will have a 1st measuring point diameter of 0.099in and a 2nd of 0.095in. All 0.125in needles have 1st measuring point diameters of 0.124in and, on average, 0.1205in 2nd measuring point diameters. These are the sizes to have present at the 1st and 2nd measuring points on each of the three common needle types in any model of carburettor.

Good idle has been achieved when the engine idles as evenly as possible and attains the required mixture strength (measured by exhaust gas C0%). Setting the idle mixture adjustment on the lean side is incorrect; the idle mixture should be set in the middle of the 'best idle' setting range. This means richening the fuel component of the idle mixture (either winding the main jet down or turning the screw on HIF carburettors) until the engine speed increases to the maximum it can for the amount of butterfly opening present at idle. The nut or screw position at this point is noted. The richness of the idle mixture is then further increased until the engine note changes and just starts to sound 'heavy' (rich). The nut or screw position is noted and the nut or screw adjusted back to approximately the middle position between the two noted points. The engine must idle without any idle misfire. Lean mixtures often result in occasional misfiring at idle speeds; richen the mixture until the idle is even.

With multiple SUs, each carburettor must have the same airflow reading at idle.

Note that once the engine idle has been set, the main jet is not touched again (raised or lowered) and all subsequent mixture changes are accomplished by altering the needle profile. This means that the 1st and 2nd measuring points are not touched during the needle reprofiling procedure. Choose needles to modify that have the listed 1st and 2nd measuring point sizes.

Check that the ignition is advanced correctly. Not having enough idle speed ignition advance can be the cause of erratic acceleration phase performance which, while being blamed on the carburettors, is actually nothing to do with them.

The area around the 2nd measuring point is the one that meters fuel for idling purposes. If this area of the needle gets polished down too much, the telltale sign is a lot of black smoke from the exhaust system when the engine is accelerated very quickly from idle when not under load (in a free revving state). When this happens make a replacement needle that is 0.001-0.002in larger at the 2nd measuring point. It's very often possible to do this without affecting the accelerative efficiency of the engine. Whatever happens, the engine must accelerate correctly: but do try and get it to do so without emitting black smoke (you may not be able to eliminate it, but you can minimise it). Too much fuel at idle can lead to sparkplug fouling problems too.

At this point the engine will be idling at a suitable engine speed, on multiple carburettor set-ups the butterflies will be set so that the airflow through each is identical at idle and the off idle (slight throttle tension being applied).

Vacuum advance

When the vacuum advance (if applicable) is reconnected after carburettor tuning the engine might develop a light misfire. If it does, check to see how much advance there is with the vacuum advance system working. If the extra advance is more than 10 degrees then this is likely to causing the misfire. It's not impossible for engines to have as much as 20 degrees of extra advance at idle, and above, when revved in unloaded condition.

If the engine runs okay under normal load, leave well alone. However, if it misses on a very light loading the solution is to limit the amount of vacuum advance possible (cut it back from 20 to 15, for example). This can be done by replacing the vacuum advance canister with one that has less advance built into it or by restricting the amount of vacuum that can get to the vacuum canister.

NEEDLE PROFILES

The next part of the process involves removing the damper and fitting a calibrated rod into the top of the carburettor. You cannot buy calibrated rods, but they are easy to make. All that's required is a lightweight rod (*e.g.* wooden dowelling, plastic or aluminium tubing) of 0.312in/7.93mm in diameter and 5-6in/127-152mm long. The rod will need to be accurately marked with SU needle measuring point positions. Each carburettor must have its own calibrated rod so that it can be clearly seen that each carburettor piston is lifting an identical amount for a given amount of throttle. The calibrated rods allow you to visually monitor the position of the rising piston and, as a consequence, needle position. The air/fuel mixture that is admitted to the engine is controlled by the height of the piston above the bridge of the carburettor body. The throttle (butterfly) can be fully open, but the air/fuel mixture is still controlled by the position of the piston.

The first measuring point is, in practice, made higher than the top of the carburettor's vacuum chamber to take into account the distance between the main jet and the top of the bridge, and the distance the stop is set proud of the base of the piston. These measurements are taken *after* the engine's idle has been set and jet heights set to optimum. With these two factors taken into account, the measuring point positions on the calibrated rods will be dead accurate. So, when the engine is not running, the 1st measuring position will be above the top of the vacuum chamber by around 0.020in/0.5mm to possibly 0.060in/1.5mm, depending on the main jet settings. When the engine is idling expect the 2nd measuring point of the needle to be near level with the top of the vacuum chamber. Idle is almost always achieved between the 1st and 2nd measuring point of the needle.

With the engine set with a good idle, the distance between the main jet and the top of the bridge can be measured. This is done by removing the vacuum chamber and piston assembly and directly measuring the distance with the tail of a vernier calliper. Whatever this dimension is, plus the thickness of the protrusion of the pad on the base of the piston, is the sum total of how much the 1st measuring point will be above the top of the vacuum chamber. This will ensure that each measuring point of the needle is positioned exactly right in relation to the calibrated rods, there will be no error.

The other SU measuring points (2nd and so on) are all marked at the prescribed 1/8in distance from each other in relation to the accurately marked 1st measuring point. When the piston is in the 'at rest' position, resting on the top of the bridge, the 1st measuring point will be, on average, anything from 0.060 to 0.090in/1.5 to 2.3mm above the top of the vacuum chamber.

You need only mark the calibrated rods with enough measuring point marks to cover full throttle (maximum piston lift). Lift the piston by hand to make sure that maximum piston travel is possible and that the last measuring point mark is visible. Extra marks beyond full throttle can lead to confusion.

Note that while all SU needles are quite long, the whole length of the needle is ***not*** used to meter fuel. This means that only those SU measuring points which are applicable to fuel metering on the particular size of carburettor need to be considered.

As the piston rises, so does the calibrated rod, and the number of marks (and part distance between marks) that are visible can be counted. When the piston is at its maximum travel (upward), the maximum number of marks will be visible. Each mark equates to the equivalent measuring point of the needle. Make up and fit identical calibrated rods to all carburettors so that each carburettor's performance can be checked.

The fact that the calibrated rods have been fitted into the top of the carburettors means that there is no damping mechanism (no carburettor will have its damping mechanism working). This means that the pistons will rise virtually instantaneously when the throttle is opened. This is of no consequence at this point as the throttle is not going to be opened quickly during the testing procedure; rather the throttle is going to be opened slowly, and then only to half throttle and not beyond.

It is important to realise that the reason SUs do not react as fast as some other carburettors is because the piston controls the rate that the air/fuel mixture is admitted to the engine. The damped, and therefore delayed, action of the rising piston causes a slightly slower engine response to throttle compared to high performance type carburettors, though overall mixture control is, of course, excellent.

For the next part of the procedure the throttle needs to be opened slowly, as slowly as you like. The engine is of course in a free revving situation. Expect engine rpm to rise as the throttle is opened but to 'hesitate' when the mixture becomes lean (if the needles are already 'right' for the application the engine will not hesitate). When the engine first 'hesitates' note the heights of the calibrated rods. Count the number of measuring point rings and partway distances exposed above the vacuum chamber top. This measuring point is where the needle is too large and needs to be polished down by 0.001-0.002in in the first instance and then the engine tested again in the same way. Keep polishing the needle (or needles) down by 0.001in or 0.002in at a time until the engine responds (doesn't falter). Naturally, all the measuring points between the modified point and the tip of the needle have to be made progressively smaller, usually by a minimum of 0.001in.

Each measuring point from the shoulder (1st measuring point) and on down to the point must be smaller than the measuring point which precedes it.

Because the standard 1st and 2nd measuring points are maintained there may well be a slight initial hesitation as the throttle is moved off the stops. This is acceptable and it is better not to start reducing the size of the 2nd measuring point to get rid of this initial leanness. It seldom shows up when driving the car. This is only allowable at the instant the throttle is opened, all other measuring points past this one must see the engine responding instantly and without faltering. If the engine 'falters,' even slightly, in this free revving state expect 'flat spot' problems on the road or track.

If, for instance, the 5th measuring point on the needle is where the engine just starts to 'falter,' the 6th measuring point will have to be made a minimum of 0.001in less in diameter, as will all subsequent measuring points down to the point.

0.001in is the minimum difference between measuring points, on a pro-rata basis, down to the point. In fact, final consecutive measuring point to measuring point diameters may range from 0.001in to 0.002in smaller than the preceding measuring point diameter. Essentially, a smaller engine will usually have smaller differences in measuring point diameter sizes compared to a larger engine which will require an overall richer mixture. A small engine is between 1000cc and 1500cc (0.001in smaller each time). A medium sized engine is 1500cc to 2000cc (0.0015in each time). A large engine is 2000cc and above (0.002in each time).

The top end mixture strength is tied into the needle point measuring sizes required to get the engine to accelerate correctly up to the mid-range. All needle measuring point sizes must get progressively smaller as they go towards the point of the needle (the actual sizes depending on the engine's requirements). With the 1st to the middle measuring point figures arrived at by the preceding testing sequence, initially make all subsequent (towards the point) measuring point sizes smaller by 0.001-0.002in: alternatively, incorporate the needle measuring point sizes from a standard needle for a similar sized engine with the same number of carburettors.

The object of this exercise is to be able to take the engine from idle through to about half full throttle (initially very slowly) without the engine 'faltering' and then to release the throttle. This means polishing the needle at a specific needle measuring point and then **all** subsequent ones towards the point of the needle, and repeating this whole process until the engine will reach half throttle without faltering.

Caution! - While the engine must be revved to test it, excessive high rpm revving (three-quarters throttle and more) is not necessary.

At this point the needle has been sized to allow the unloaded engine to go from idle to half throttle very slowly without faltering. Unfortunately, all of the needle diameters will have to be reduced further to give correct mixture strength for a loaded engine with carburettor dampers fitted. However, the overall profile of the needle has now been established and this will not subsequently change.

The throttle is not 'snapped' open in the following tests, rather it is opened quickly (there is a difference). A throttle that is 'snapped open' is opened virtually instantly, while a throttle which is opened quickly is opened to the half full throttle point in about one second.

Essentially, the process already described for slow throttle opening and needle polishing at the measuring point which sees the engine faltering is repeated, but now with the throttle opened quickly. Start polishing the needle at the measuring point where the engine falters, and then polish progressively down to the needle tip to maintain a minimum of 0.001in difference between measuring points. It's fair to say that these recommended size differences may end up being too great (needle becomes too small in places) and the top end fuel mixture ending up too rich for the engine. New needles will have to be made which are larger at these particular measuring points. An engine which is run with an excessively rich mixture at the top end will be slower than it would be with the correct mixture.

Re-test the engine after each needle alteration and expect the engine to rev better at each re-test.

When the engine can be free revved cleanly (without any faltering) from idle to about half throttle in about one second, the air/fuel mixture will be about right for driving under load (believe it or not). When the throttle action is this quick and there is definitely no hesitation of any description (not a hint) the needle is removed from the carburettor and all sizes past the last polished measuring point checked to make sure that they are all smaller, by a minimum of 0.001in, than the last measuring point worked on.

To check that the top end mixture is sufficient to run the engine under load, the throttle will have to be 'snapped' open quickly to the three-quarter full throttle point to see whether the engine falters. This is not really a very good thing to do to an engine, but it is necessary in this instance. If the engine rpm rises immediately without a hint of faltering, do not let the engine rev out fully as that is not the object of the exercise. What you are looking for is the willingness of the engine to rev when the butterfly is opened quickly beyond half throttle. This process normally takes about half a second and does the engine no harm - provided the revs are not allowed to climb too high. If the revs immediately start to go 'sky high,' close the throttle immediately as the chances are that the needle is near right.

If the engine falters under the snap acceleration test, polish all measuring points from the last measuring point worked on, and down to the point of the needle, by a further 0.001in. Try the test again to check the mixture. Keep polishing the needle measuring point at which faltering occurs (and subsequent points) until the engine revs cleanly.

When the mixture is too lean the engine will not rev cleanly as not enough fuel is being supplied. When the mixture is rich enough the engine will rev instantaneously and cleanly. An over-lean mixture will cause the engine to hesitate and likely backfire through the carburettor. With an over-rich mixture the engine will rev, but will sound 'heavy.'

You have now reached a point where the engine is basically ready for limited road/track/dyno testing. Ultimately, the needles may need further alteration but the engine will now be usable. There is still further testing to be done to finalise the needle measuring point sizes at the top end, but from now on the engine will be tested under load. The damper/s can be fitted to the carburettor/s and oil added to the dashpot/s.

The engine can now be tested up to half full throttle on the road/track/dyno to see how it actually goes, and whether or not there is any hesitation at all in the low to mid rpm range. The engine is checked at progressively higher revs to avoid possible damage from a lean mixture condition.

Initially use up to half throttle to go through the gears and depress the accelerator pedal as quickly as you like. If there is no hesitation when using up to half throttle, use three-quarter throttle to accelerate through the gears and finally full throttle.

If there is any hesitation (this would be unusual when the needle measuring points have been arrived at by the previously described method, but not impossible), go back and repeat the previous procedure from the slow throttle test to the snapped open throttle test. If there is even slight hesitation somewhere in the rpm range, it will be enough to cause a flat spot under load. Polish the needle/s (0.001in at a time) at the relevant measuring point until there is absolutely no hint of hesitation. The rest of the needle down to the point will have to be repolished (resized), so that each successive measuring point is smaller than the one that precedes it.

Hesitation during acceleration almost always means a lean mixture somewhere in the range of needle movement. **Caution!** - Because an engine can be damaged by a lean mixture, care must be taken to ensure that the engine is never run too lean. 5% CO is the full power mixture (exhaust gas analysed) required. Suitable measuring equipment is available on rolling road dynos. Alternatively, a sparkplug check can be done. This means running the engine (at full operating temperature) at maximum rpm under load in one of the higher gears for at least a mile, cutting it dead and removing the sparkplugs: you then check the colour of the insulators. A white insulator means a lean mixture, while a chocolate brown colour means a correct type of mixture.

Having a range of needles made up is the best way of obtaining maximum power and avoiding engine damage. Needles can be made up that are richer than the ones you've already made to see if the engine goes better with these. Time the car over a measured distance, using top gear only from mid range to maximum rpm, with each type of needle fitted to the carburettors. Use rich needles first (smaller diameters at the measuring points). Start rich and then go leaner, not the other way around. Avoid prolonged high rpm until you are sure that the top end fuel/air mixture is right. If the engine goes better with one rather than another needle, this is a very good indication that the mixture offered by that needle is better than the other. To get this sort of effect there may only be a 0.001-0.002in difference in the higher needle measuring point diameters. It can take as little as this to make the difference between the engine being slower or faster (and one that stays in one piece). When the engine is used hard, black smoke from the exhaust is an indication of an over-rich mixture, as is a jet black exhaust tailpipe interior (dark grey is the correct colour). The sparkplugs tell it all though, provided the engine is in perfect mechanical condition.

In some instances, having a 0.001in minimum difference in needle diameter between measuring points may mean that the top end mixture is too rich. If this is the case, reduce the minimum difference between consecutive measuring points to 0.0005in. Note that this small amount of diameter size difference at consecutive measuring points will only apply between the 7th measuring point and the point of the needle for most needles.

If you record all needle measuring point sizes onto a diagram, the exact nature of the richness/leanness factor will be evident. This information will help you to arrive at alternative profiles which are ideal for your application.

Chapter 10
Modifications for racing engines

All previous text in this book deals with standard carburettors and the fitting of various needles (maybe modified) and standard piston springs as made by SU (Burlen Fuel Systems). This chapter is for those who wish to do a bit more to their carburettors besides swapping needles and piston springs to get the best possible engine performance from the carburettor for racing purposes.

Ideally, the spindle modifications suggested are required, but are not *absolutely* necessary. Top end power is not always improved all that much (but usually is) by the spindle modifications; however, the acceleration of the engine is almost always improved - and that's usually what matters most. The result of the spindle modifications is that the carburettor is, effectively, a little bit bigger (greater cross-sectional area) than standard and, further to this, the throttle bore is more streamlined: these two things add up to easier airflow and more airflow. Because of the amount of work involved, it is recommended that throttle spindles be left as standard in the first instance: they can always be modified at a later stage, if required (needles may need altering too). If the spindle modifications are attempted, but you are not able to complete the work successfully, at worst you'll have to buy a new spindle kit.

The modifications detailed in this chapter result in an improvement in engine performance (acceleration, rather than a dramatic increase in maximum power), but they also result in higher fuel consumption (30%, or more) and the engine will often be 'fussy' just off idle, especially if a long duration camshaft is fitted. This 'fussiness' off idle is caused by the carburettor piston 'bouncing' (rising and falling rapidly, often audibly) due to the erratic inlet tract airflow at low engine speeds.

Realistically, these modifications are only for competition purposes and are not recommended for road going cars because there is likely to be some off idle hesitation. If a competition engine suffers slight hesitation coming off idle it seldom causes problems. Many competition engines idle at 1200-1500rpm and don't come 'on cam' until 3500-4000rpm making the lower rev range largely irrelevant. However, engines fitted with long duration camshafts can be tamed by fitting very light piston springs and making needles that have needle measuring point sizes at the 1st, 2nd and 3rd points that are rich, but not so rich as to cause plug fouling. The piston spring should be light enough not to interfere with rapid lifting of the piston during acceleration, while still offering reasonable resistance to the piston to reduce 'piston bouncing.' This solution does of course represent a slight compromise because it would be better if the piston spring was not used at all.

While the modifications detailed in this chapter will often work well on a road car, the loss of fuel economy, the increase in idle speed exhaust emissions and the prospect of possible fuel fouled sparkplugs in traffic is against it. In a road going situation you are generally better off setting the carburettors up with optimised needles, standard light piston springs and having oil in the dashpot.

It doesn't matter how many SUs there are on the engine, they can all be given the same treatment. In many instances the results will be as good, or nearly as good, as the same engine equipped with Weber or Dellorto carburettors. This may sound a bit far-fetched, but it isn't.

SUs fitted to a road car deliver near unmatched fuel economy with reasonable power (as good, or better than, most other mass production-type carburettors). The original SU carburettor was, of course, designed to deliver good all round performance as economically as possible and, in standard form, it does just that. Essentially the listed 'modifications' maximise power without considering economy at all. The good fuel economy factor is lost through the modifications as is, to some degree, low end tractability, but these adverse aspects can be reversed in about two minutes per carburettor. This means that a car could be driven to the race track, have the changes made, race all day with the carburettors in modified form and then, after changing the carburettors back to standard form, be driven home with good fuel economy.

The main reason that SUs do not have a high performance image is because they cannot provide immediate engine acceleration when the throttle is opened quickly, unlike sidedraught Webers/Dellortos for example. The underlying reason for this is the fact that the SU piston is slow to rise and admit more air/fuel to the engine. There is no accelerator pump on an SU and this does create a slight problem: rising of the piston has to be delayed to avoid leaning the mixture off too much when the throttle is opened quickly (piston movement is resisted by the spring and the oil in the dashpot), and the needle is essentially

Standard SU carburettor spindle.

Modified SU carburettor spindle with the standard butterfly screws fitted and ready to be ground down.

sized to suit piston spring pressure. Other types of carburettor use an accelerator pump to 'cover' the throttle being snapped open or opened quickly. If the piston of an SU is to rise quickly the amount of fuel being admitted to the engine simply has to be increased to prevent hesitation.

The SU's slight acceleration delay is not all that noticeable in a normal road going situation. It is however noticeable in a racing situation, when

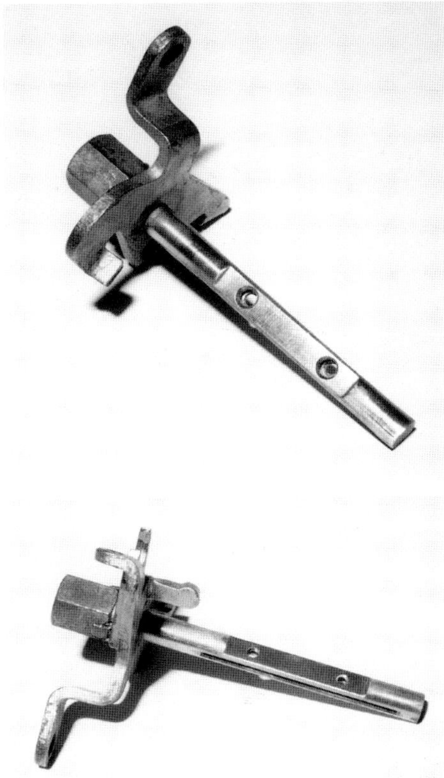

Both sides of the spindle are reduced, not just one.

Thinned down spindle after screws have been ground down and the spindle 'flash soldered' (see text).

other competitors are equipped with Webers or Dellortos; the usual consequence is that SUs get replaced by 'better' carburettors. This scenario changes when the SU is modified and

44mm HIF with an altered spindle shown at full throttle.

44HIF spindle and screws flash soldered.

there is often virtually no difference in performance to that of Weber or Dellorto equipped engines - unless the SUs are actually too small (not enough airflow capability) or, for that matter, too large. Make no mistake about it, the SU can be made to deliver absolutely brilliant power and do what it was not really designed to do, which is to compete virtually head-on with the best high performance-type carburettors available. However, this can only happen once the piston spring and damper oil are removed.

THE MODIFICATIONS

Although the modifications can involve streamlining (thinning) the carburettor throttle spindle, this is definitely an option. The main part of the modification process is the removal of the piston spring from the vacuum chamber, not using any oil in the dashpot (save for a little lubrication) and fitting specially made needles.

New needles will almost certainly have to be made: alter original needles to optimal measuring point sizes found by using the calibrated rod system.

Important -Before starting to convert your engine to run with modified carburettors it must be running perfectly well. This means that multiple carburettors must be synchronised and all carburettors must have oil in the dashpots, piston springs fitted to the vacuum chambers and the needles must be 'right' all through the rpm range (no flat spots). The modifications detailed in this chapter are better carried out as a progression from this situation.

Throttle spindle modifications

As can be seen from the photos, the spindle modifications should improve carburettor airflow, but be aware that these modifications don't always lead to a measurable improvement in engine performance.

A 1 1/4in SU's 0.25in diameter spindle can be reduced to 0.15in thickness. Other SUs have 0.312in spindles, which can be narrowed down to 0.18in in thickness. The original butterfly screws can be put back in, tightened and the protruding parts above the spindle ground away so that the surface is flush. The area where the protruding heads of the screws were, needs to be soldered over with a thin 'flash' of solder to prevent the screws from coming undone. **Caution!** - don't rely on a locking agent alone to hold the screws in place. There's a product on the market called "Solder Paint" which comes in paste form. To 'flash coat' the screws, daub each with a little Solder Paint and then heat the spindle with a propane gas flame - it does not take much heat to cause the solder to flash (liquify).

Needles will almost certainly have to be made (altered standard needles actually) as you cannot normally buy standard needles to suit carburettors modified in this manner. Some part of a standard needle will almost certainly have to be reduced in size. That is not to say that a particular standard needle may not suit a particular application, just that it is not likely in most instances. Smaller engines are most likely to be able to use needles as made by SU (the needles will effectively be rich needles made for larger capacity engines). Making needles (altering standard needles) is not all that difficult if the given procedure is followed. It does take some dexterity to alter needles and practice makes perfect. Expect to ruin the odd needle or two.

Normally if the spring and recommended weight oil are left out of the carburettors the engine simply will not accelerate at all (as in being undriveable). This would be a completely hopeless situation, except for one thing: the fact that a needle can be sized to provide enough fuel to cover for a very rapidly rising piston. This does mean that the air/fuel mixture may well end up being slightly rich low down, but this factor often applies to other high performance carburettors too. The air/fuel mixture throughout the range will be in the region of 11.8:1 or 12.5:1 (5%-6% CO) which is a full power mixture.

Unfortunately, it is sometimes a bit difficult to get the needle sizing right for low down engine performance (because of the absence of a piston spring). On a racing engine this does not normally matter as, once the throttle is opened and the revs rise the engine is over the initial acceleration phase. To overcome the slight off idle

Remove a minimum of 0.060in to a maximum of 0.100in from the two faces arrowed.

Fibre washer is 0.020in thick and, if necessary, you can use two. The washer goes over the spindle (arrowed) of the piston to limit its upward travel.

hesitation problem, the needle may have to be made excessively rich, which could result in sparkplug fouling at low engine speeds - a problem which can be very annoying.

If clean pick-up just off idle is essential, set up the carburettor as described and then fit a very light piston spring. The spring should exert just enough pressure to make sure than the piston is under tension (minimal) when resting on the bridge of the carburettor (engine not running). This work-around is often enough to smooth out off idle fussiness. An alternative is to use a very light oil in the dashpot (3 in 1 oil, for example). Try both methods to see which works best with your application.

Dashpot cap vent hole modification

The vent hole in the top of the dashpot is approximately 0.040in/1mm in diameter and this diameter is always sufficient for any SU carburettor fitted with a piston spring and with oil in the dashpot. However, when these carburettors are modified in the manner suggested, there can sometimes be a noticeable improvement in the rate of power delivery if the vent hole is increased in diameter to 1/8in/3.2mm. This allows faster expulsion of air and the piston to rise faster and allow 'cleaner' acceleration. Always consider this vent hole modification if there is any slight hesitation under maximum acceleration.

If there is no improvement in acceleration after the vent hole has been enlarged, the dashpot caps do not need to be renewed as the larger hole is not in anyway detrimental to performance. Many engines which have these modifications can retain the original air cleaner system, as opposed to running 'ram tubes' in the open air as is often done with a high performance engine. The original air cleaner system (paper air filter) will almost always prove easier to finalise the needle sizes (low, mid and top end performance will all be good). The standard air filter capacity is frequently quite sufficient, and it is just so convenient to retain the original equipment (can't have it less expensive than that).

Air cleaners/ram tubes

Many people like to fit 'ram tubes' or trumpets to SUs and run them in the open air, even though this means unfiltered air is going into the engine (expect high engine wear). In reality the low end performance with ram tubes fitted will often not be good (fussy) compared to the very same engine running the original air cleaner system. Ram tubes really need to be enclosed in a reasonably large air box which has as large a paper element air filter as possible incorporated. No engine should have unfiltered air going into it: engines do not normally lose power through having good quality paper air filtration, they lose power if the filter is too small (insufficient airflow capacity) or is dirty. Having the original air cleaner system fitted might not look too 'sporty,' but the results are what matter in the final analysis. Virtually no engine has an inefficient

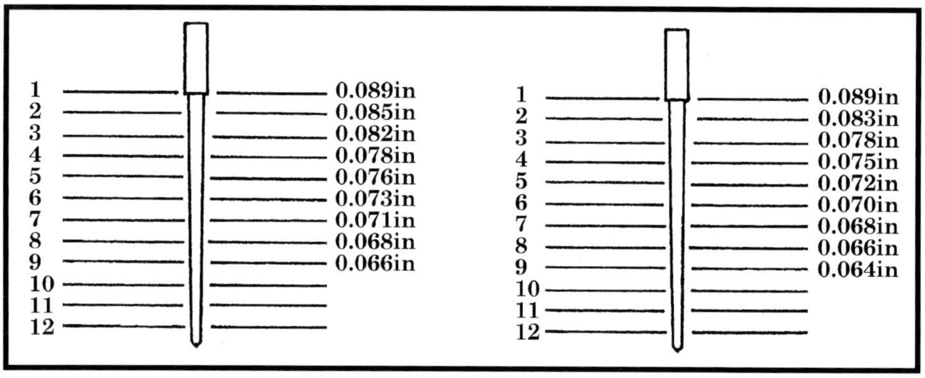

Needle on the right is for a modified carburettor, while the needle on the left is for an unmodified carburettor. The needles shown are for 1 1/4in SUs and the measuring points go down to the piston full lift position only.

standard equipment paper air filtration system fitted. That goes for single or multiple carburettor installations too.

Of course, it is possible that the original air cleaner system will become restrictive after a certain level of airflow is exceeded. This poses a bit of a problem insofar as how do you tell whether the system is restrictive, or not? This can be determined by running the engine without an air cleaner system for the shortest possible time to see if performance improves. If the original system does prove restrictive, investigate the many aftermarket air filtration systems available. You may find that some additional work on the needles may be needed to optimise performance with non-standard air cleaners fitted.

Note that Burlen Fuel Systems (SU) do not sanction the modifications suggested here in any way. However, they will supply any parts and are well aware that many people are looking for maximum performance from SU carburettors in certain applications.

Checking/modifying needles for correct mixture strength

Procedure described in chapters 8 & 9.

Piston travel (1 1/4in SU)

The 1 1/4in SU has the piston travel limited by design so that the piston does not go all the way to the top of the throttle bore. (A 1 1/4 SU with a modified spindle and increased piston travel flows nearly as much air as a 1 1/2 SU.) The piston would need to go up a further 0.100in to reach the top of the throttle bore diameter. Once the throttle spindle has been streamlined the piston should be allowed to rise a minimum of a further 0.060in. This is achieved by machining down the top of the piston spindle by 0.060in and the base of the central pillar or bush of the vacuum chamber by 0.60in. Better still is to machine 0.100in off these two surfaces which means the piston can rise 0.100in higher than a standard carburettor.

If the carburettor is tested with the 0.100in modification and you think that perhaps the top end performance is not quite right, place a 0.020in thick fibre washer over the piston spindle. This will limit the travel of the piston to 0.080in over standard as opposed to 0.100in. Use all of the travel if the engine responds, cut it back by using washers if the engine does not respond.

The needle of an SU modified in the foregoing manner will almost always have a different overall profile compared to the needle which would be ideal in the same carburettor without these modifications. The 1st measuring point will be the standard SU size (0.089, 0.099 or 0.124in), while the second measuring point will likely be slightly smaller by 0.002-0.003in but, after this, the needle will reduce in diameter dramatically to provide a rich mixture from just off idle to mid-range. From the three-quarter piston lift measurement point downwards, the needle will be sized much the same as any needle designed to be used with an unmodified carburettor on the particular engine.

The secret of getting SUs to work without a piston spring and damper oil is to make a suitable needle. If the needle's not right, it won't work.

Manual choke

The choke mechanism will have to be operational to start the engine. The usual way of setting an engine that has had the carburettor modified as described here is to set the choke so that the throttle part of the combination is not activated. This means that when the choke is pulled out the accelerator pedal is used to set the engine rpm (not the cam as fitted to the carburettor), and the use of the choke only enriches the fuel mixture (simply lowers the main jet). By all means set the choke so that it does work as per normal, if preferred. For a racing engine, expect it not to need much, if any, enriched mixture once it has fired and has been running for more than 10 seconds.

The choke will not normally work successfully if the throttle component (butterfly) is in operation before the main jet starts to lower. Set the choke so that the instant the main jet starts to move, the butterfly starts to open as well (as per standard).

Chapter 11
Your needle modification record

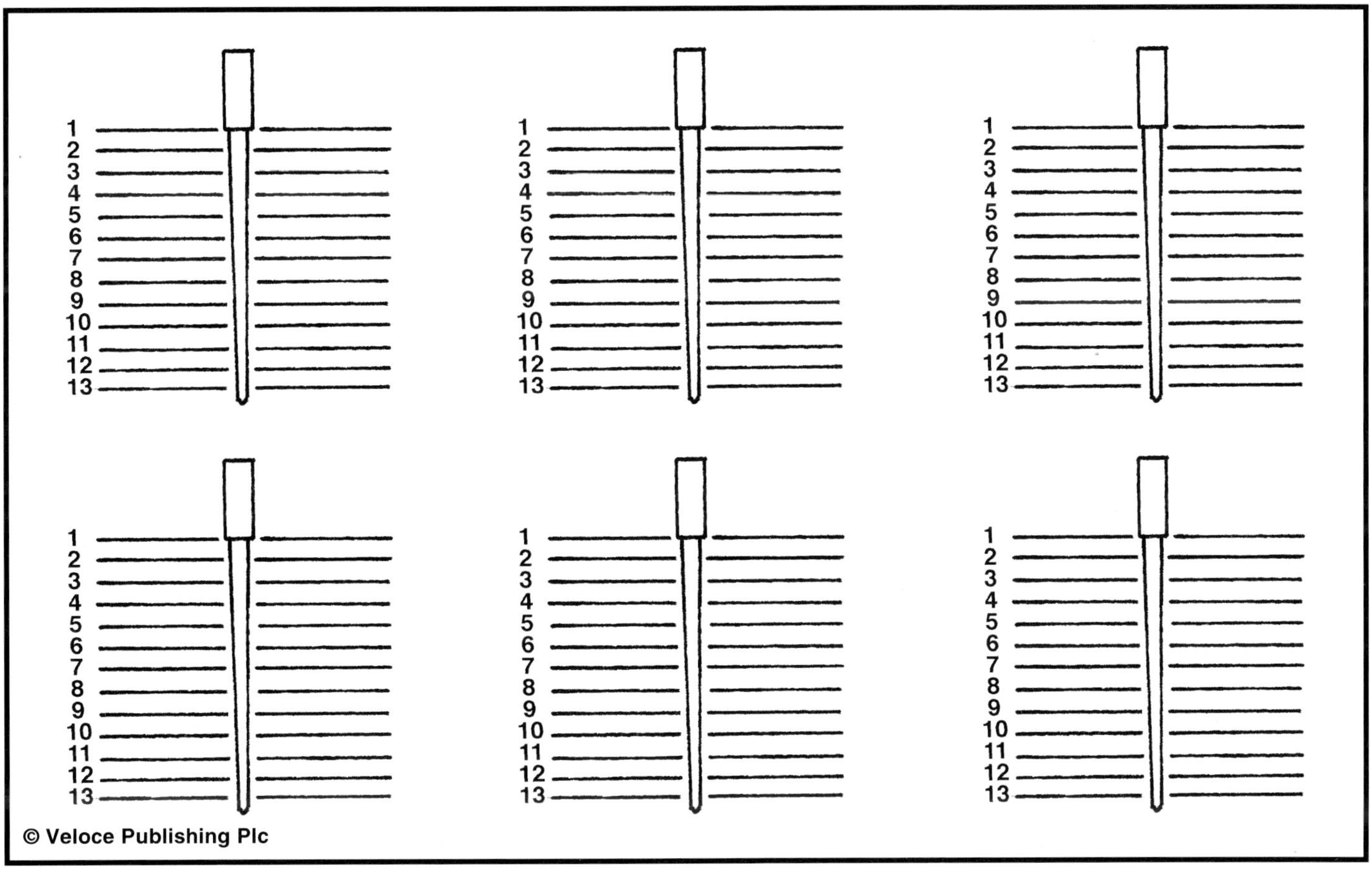

0.90in Fixed-type SU needle.

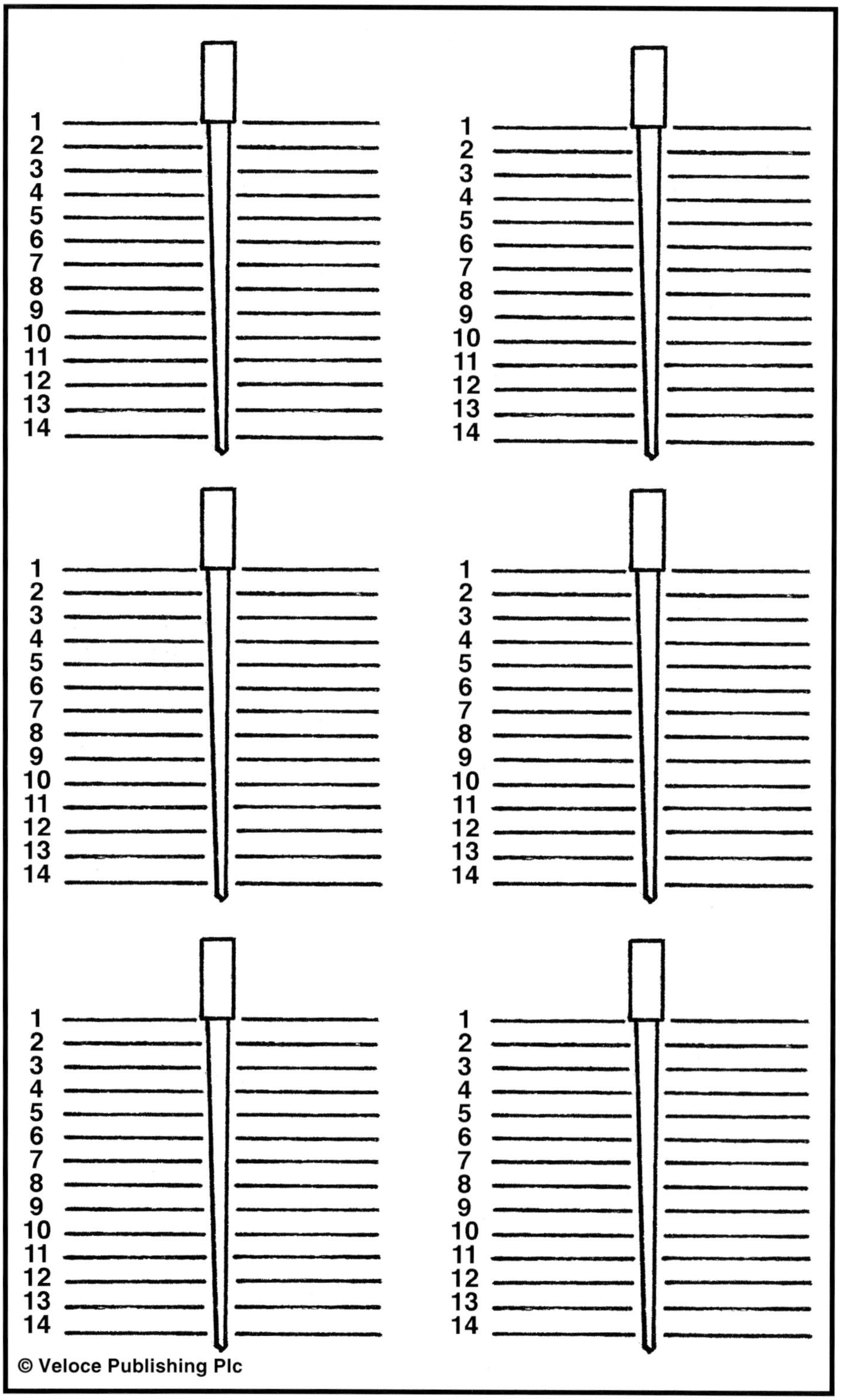

0.100in fixed-type SU needle.

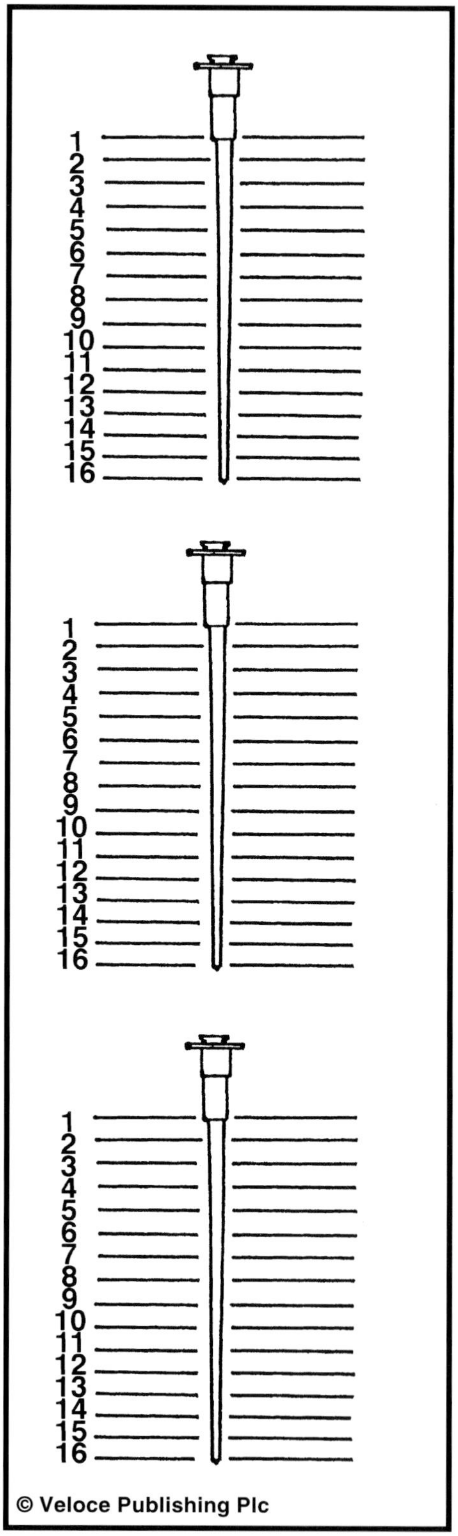

0.090in & 0.100in spring-loaded needle.

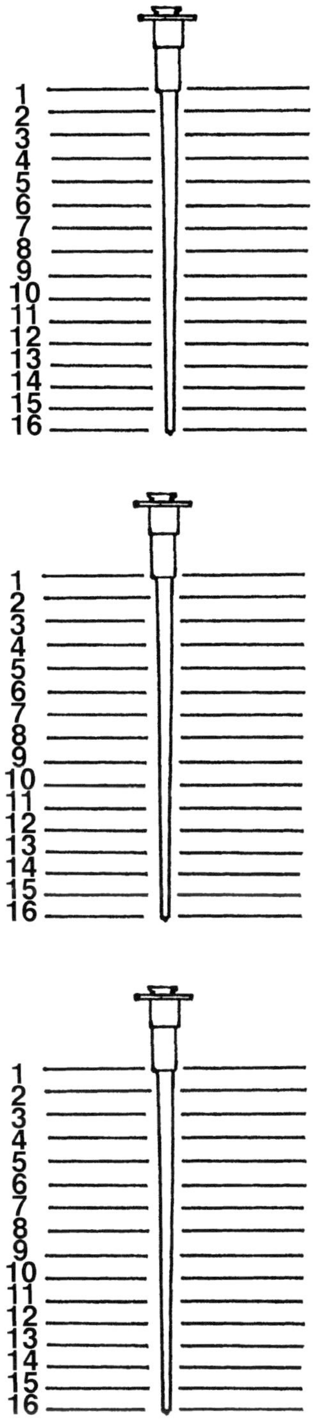

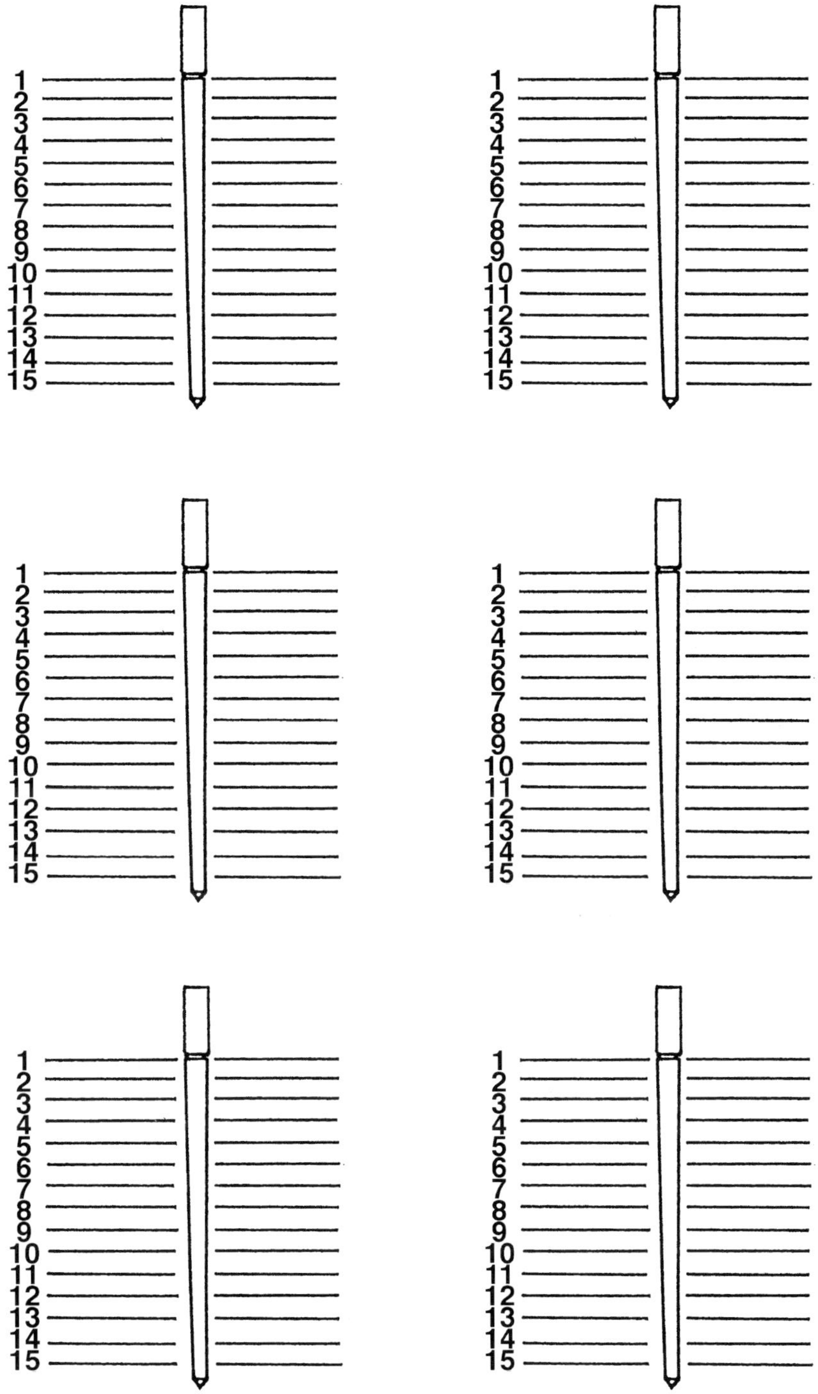

0.090in & 0.100in spring-loaded needle.

2in SU fixed-type needle.

ALSO FROM VELOCE PUBLISHING -

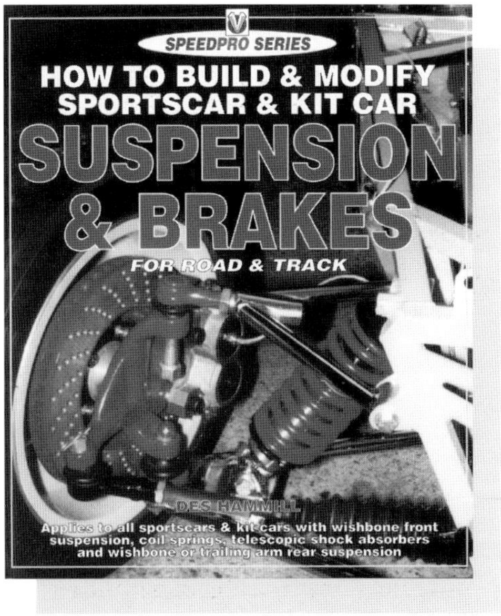

HOW TO BUILD & MODIFY SPORTSCAR & KIT CAR SUSPENSION & BRAKES

by Des Hammill

ISBN 1 901295 08 7
Price £14.99*

A book in the ***SpeedPro*** series.

- Applies to all two seater sportscars and kit cars with wishbone front suspension, coil springs, telescopic shock absorbers and wishbone or trailing arm rear suspension.
- Basic information applies to all cars.
- Written in a clear understandable style with over 100 detailed original diagrams.
- Cuts through the mystique and confusion surrounding suspension and handling improvements.
- Ideal for the home mechanic.
- Applies to road and track applications.
- Des Hammill is an engineer and a professional race car builder with many years of practical experience.

CONTENTS

Chassis integrity • Suspension geometry • Ride height • Negative camber, castor and Kingpin inclination • Springs and shock absorbers • Brakes • Setting up the car • Testing, alterations and adjustments • Index

SPECIFICATION

Paperback. 250 x 207mm (portrait). 112 pages. Over 100 detailed original diagrams.

RETAIL SALES

Veloce books are stocked by or can be ordered from bookshops and specialist mail order companies. In case of difficulty we can supply direct (credit cards accepted).

* *Price subject to change.*

Veloce Publishing Plc, 33 Trinity Street, Dorchester, Dorset DT1 1TT, England. Tel: 01305 260068/ Fax: 01305 268864/E-mail: veloce@veloce.co.uk

Visit Veloce on the Web - www.veloce.co.uk

Index